METHODS IN MOLECULAR BIOLOGY

Series Editor
John M. Walker
School of Life Sciences
University of Hertfordshire
Hatfield, Hertfordshire, AL10 9AB, UK

For further volumes:
http://www.springer.com/series/7651

Nuclear Reprogramming

Methods and Protocols

Second Edition

Edited by

Nathalie Beaujean

INRA, Jouy-en-Josas, France

Hélène Jammes

INRA, Jouy-en-Josas, France

Alice Jouneau

INRA, Jouy-en-Josas, France

 Humana Press

Editors
Nathalie Beaujean
INRA
Jouy-en-Josas
France

Hélène Jammes
INRA
Jouy-en-Josas
France

Alice Jouneau
INRA
Jouy-en-Josas
France

ISSN 1064-3745   ISSN 1940-6029 (electronic)
ISBN 978-1-4939-1593-4   ISBN 978-1-4939-1594-1 (eBook)
DOI 10.1007/978-1-4939-1594-1
Springer New York Heidelberg Dordrecht London

Library of Congress Control Number: 2014950228

Humana Press is a brand of Springer
Springer is part of Springer Science+Business Media (www.springer.com)

Preface

Somatic cell nuclear transfer (SCNT) cloning is a unique technology able to reprogram the somatic cell genome into a state equivalent to that of the fertilized oocyte: the so-called totipotent state. This technique can be used to produce an animal from a single cell nucleus using an enucleated oocyte as a recipient. Since the birth of Dolly, a cloned sheep from adult somatic cells, many attempts have been made to clone animals of different species with various donor cell types, including mouse, rabbit, pig, and cattle. SCNT technology is expected to be useful for farm animal breeding (cloning of elite animals), medicine, and pharmaceutical manufacturing with the production of transgenic animals and conservation of endangered species. Although the obtention of iPS cells seems promising in some domestic species, generating transgenic animals would remain a long and expensive process, especially in species with singleton litter. Therefore, nuclear transfer remains the best means to obtain transgenic animals. With the new technologies in genome editing such as Talen, Zinc-finger nucleases, or CRISP/CAS, precisely engineered transgenic cells can be obtained, which then can be used as donor cells.

To date, the efficiency of SCNT is low, despite the numerous studies that have been performed. Losses of the cloned embryos not only occur throughout early preimplantation stages but also during postimplantation and pre- and postnatal development with rates significantly lower than those resulting from in vitro-produced embryos.

Based on the accumulated information on the various worldwide studies, it is broadly accepted that the SCNT efficiency can be affected by a number of biological and technical factors. However, in the past 10 years, many technical improvements have been made to improve SCNT efficiency, with a special emphasis on epigenetic approaches. Indeed, enhancement of the levels of histone acetylation in the chromatin of the reconstructed embryos can efficiently help nuclear reprogramming and enhance the production of SCNT offspring. Thanks to these recent progresses and although it will take a considerable time before we fully understand the nature of genomic programming and totipotency, we may expect that somatic cell cloning technology will soon become broadly applicable.

Besides these practical applications, SCNT is also a unique opportunity for genomic research, especially in epigenetics, to learn how the somatic cell genome is reprogrammed into totipotency. This is also an irreplaceable tool to study the epigenome stability over time and between genetically identical animals. Moreover, compared to reprogramming induced by forced expression of transcription factors, nuclear transfer is an efficient means to reprogram somatic cells into pluripotent cells. Whereas the obtention of induced pluripotent stem (iPS) cells takes at least 4 weeks and dozens of cell divisions, the formation of the pluripotent cells in the embryo reconstructed by nuclear transfer takes only 4 days in the mouse (time to develop up to blastocyst stage) and 6–7 cell divisions. Embryonic stem (ES) cells have been derived from cloned blastocysts and have been shown to be very similar to ES cells derived from fertilized embryos. This is exemplified by their successful use in diploid chimera formation and tetraploid complementation.

In this revised edition of Nuclear Reprogramming, we have included not only the classic methods to perform nuclear transfer in different species but also methods that are technically less demanding while allowing the study of the early steps of reprogramming, such as heterokaryon or nuclear transfer into Xenopus oocyte. We also provided several techniques to assess the early and late development of the reconstructed embryos, at the cellular, molecular, and epigenetic level. Therefore we hope this edition will interest not only cloners but also researchers interested in studying the development of reconstructed embryos up to the adult stage.

We are grateful to all scientists who contributed to the chapters of this book.

Jouy-en-Josas, France   *Nathalie Beaujean*
                                                                            Hélène Jammes
                                                                            Alice Jouneau

Contents

Contributors

SEBASTIAN T. BALBACH • *Max-Planck Institute for Molecular Biomedicine, Münster, Germany*

NATHALIE BEAUJEAN • *UMR1198, Biologie du Développement et Reproduction, INRA, Jouy-en-Josas, France*

BERTRAND BED'HOM • *UMR1313, Génétique Animale et Biologie Intégrative, INRA, Jouy-en-Josas, France; UMR1313, Génétique Animale et Biologie Intégrative, AgroParisTech, Paris, France*

THIERRY BLACHÈRE • *Institut Cellule Souche et Cerveau, INSERM U846, Bron, France*

MICHELE BOIANI • *Max-Planck Institute for Molecular Biomedicine, Münster, Germany*

AMÉLIE BONNET-GARNIER • *UMR1198, Biologie du Développement et Reproduction, INRA, Jouy-en-Josas, France; ENVA, Maisons Alfort, France*

CLAIRE BOULESTEIX • *UMR1198, Biologie du Développement et Reproduction, INRA, Jouy-en-Josas, France*

JANICE BRITTON-DAVIDIAN • *Institut des Sciences de l'Evolution de Montpellier, Université Montpellier II, Montpellier, France*

VINCENT BROCHARD • *UMR1198, Biologie du Développement et Reproduction, INRA, Jouy-en-Josas, France*

SABINE CHAUVEAU • *GReD, INSERM 1103, UMR CNRS 6293, Clermont Université, Clermont-Ferrand, France*

CLAIRE CHAZAUD • *GReD, INSERM 1103, UMR CNRS 6293, Clermont Université, Clermont-Ferrand, France*

PATRICK CHESNÉ • *UMR1198, Biologie du Développement et Reproduction, INRA, Jouy-en-Josas, France*

JOHN ARNE DAHL • *Division of Diagnostics and Intervention, Department of Microbiology, Institute of Clinical Medicine, Oslo University Hospital, Rikshospitalet, Oslo, Norway*

NATHALIE DANIEL • *UMR1198, Biologie du Développement et Reproduction, INRA, Jouy-en-Josas, France*

ISABELLE DUFORT • *Département des Sciences Animales, INAF, Faculté des Sciences de L'agriculture et de L'alimentation, Centre de Recherche en Biologie de la Reproduction, Université Laval, Québec, QC, Canada*

VÉRONIQUE DURANTHON • *UMR1198, Biologie du Développement et Reproduction, INRA, Jouy-en-Josas, France*

HÉLÈNE HAYES • *UMR1313, Génétique Animale et Biologie Intégrative, INRA, Jouy-en-Josas, France; UMR1313, Génétique Animale et Biologie Intégrative, AgroParisTech, Paris, France*

HANNE M. HOLM • *Department of Clinical Veterinary and Animal Sciences, University of Copenhagen, Copenhagen, Denmark*

POUL HYTTEL • *Department of Clinical Veterinary and Animal Sciences, University of Copenhagen, Copenhagen, Denmark*

ALICE JOUNEAU • *UMR1198, Biologie du Développement et Reproduction, INRA, Jouy-en-Josas, France*

JEROME JULLIEN • *Wellcome Trust/Cancer Research UK Gurdon Institute, University of Cambridge, Cambridge, UK*

BARBARA KESSLER • *Laboratory for Functional Genome Analysis (LAFUGA), Gene Center, Ludwig-Maximilians-Universität München, Munich, Germany*

DAULAT RAHEEM KHAN • *Centre de Recherche en Biologie de la Reproduction (CRBR), Université Laval, Québec, QC, Canada*

HÉLÈNE KIEFER • *UMR1198, Biologie du Développement et Reproduction, INRA, Jouy-en-Josas, France*

HIROSHI KIMURA • *Graduate School of Frontier Biosciences, Osaka University, Suita, Osaka, Japan*

ARNE KLUNGLAND • *Clinic for Diagnostics and Intervention, Institute of Medical Microbiology, BIG CAS-OSLO Genome Research Cooperation, Oslo University Hospital, Rikshospitalet, Oslo, Norway*

MAYUKO KUROME • *Laboratory for Functional Genome Analysis (LAFUGA), Gene Center, Ludwig-Maximilians-Universität München, Munich, Germany*

JOON-HEE LEE • *Department of Animal Bioscience, College of Agriculture and Life Sciences, Institute of Agriculture & Life Science, Gyeongsang National University, Jinju, South Korea*

ANNICK LEFÈVRE • *Institut Cellule Souche et Cerveau, INSERM U846, Bron, France*

WEI LI • *State Key Laboratory of Reproductive Biology, Institute of Zoology, Chinese Academy of Sciences, Beijing, China*

XIN LI • *State Key Laboratory of Reproductive Biology, Institute of Zoology, Chinese Academy of Sciences, Beijing, China*

ZICHUAN LIU • *UMR1198 Biologie du Développement et Reproduction, INRA, Jouy-en-Josas, France; Friedrich Miescher Institute for Biomedical Research, Basel, Switzerland*

WALID E. MAALOUF • *Faculty of Medicine & Health Sciences, The University of Nottingham, Nottingham, UK*

EIJI MIZUTANI • *University of Yamanashi, Kofu, Yamanashi, Japan*

HIROSHI NAGASHIMA • *Meiji University International Institute for Bio-Resource Research, Kawasaki, Japan*

OLGA ØSTRUP • *Department of Tumor Biology, Oslo University Hospital RH, Oslo, Norway*

HANNE S. PEDERSEN • *Department of Animal Science, Aarhus University, Aarhus, Denmark*

NATHALIE PEYNOT • *UMR1198, Biologie du Développement et Reproduction, INRA, Jouy-en-Josas, France*

BERND PÜSCHEL • *Institute of Anatomy and Embryology, University Medical Center, Göttingen, Germany*

CLAUDE ROBERT • *Département des Sciences Animales, FSAA, Centre de Recherche en Biologie de la Reproduction, Université Laval, Québec, QC, Canada*

MARC-ANDRÉ SIRARD • *Département des Sciences Animales, INAF, Faculté des Sciences de L'agriculture et de L'alimentation, Centre de Recherche en Biologie de la Reproduction, Université Laval, Québec, QC, Canada*

HUSEYIN SUMER • *Monash Institute of Medical Research, Monash University, Clayton, VIC, Australia*

ANNE-CLÉMENCE VEILLARD • *UMR1198, Biologie du Développement et Reproduction, INRA, Jouy-en-Josas, France*

PAUL J. VERMA • *Monash Institute of Medical Research, Monash University, Clayton, VIC, Australia; South Australian Research Institute (SARDI), Rosedale, SA, Australia*

SAYAKA WAKAYAMA • *University of Yamanashi, Kofu, Yamanashi, Japan*

TERUHIKO WAKAYAMA • *University of Yamanashi, Kofu, Yamanashi, Japan*

ECKHARD WOLF • *Laboratory for Functional Genome Analysis (LAFUGA), Gene Center, Ludwig-Maximilians-Universität München, Munich, Germany*

ANNEGRET WUENSCH • *Laboratory for Functional Genome Analysis (LAFUGA), Gene Center, Ludwig-Maximilians-Universität München, Munich, Germany*

BAO-LONG XIA • *State Key Laboratory of Reproductive Biology, Institute of Zoology, Chinese Academy of Sciences, Beijing, China*

KAZUO YAMAGATA • *Laboratory for Genomic Reprogramming, Center for Developmental Biology, RIKEN, Kobe, Japan; Center for Genetic Analysis of Biological Responses, Research Institute for Microbial Diseases, Osaka University, Suita, Osaka, Japan*

QI ZHOU • *State Key Laboratory of Reproductive Biology, Institute of Zoology, Chinese Academy of Sciences, Beijing, China*

Chapter 1

Nuclear Transfer in the Mouse

Vincent Brochard and Zichuan Liu

Abstract

Nuclear transfer (NT) technique provides a powerful experimental tool to study the mechanisms of reprogramming processes and to derive NT-embryonic stem (ntES) cells from living or frozen animals. The Piezo-driven direct microinjection NT method has proved to be a valid technique to clone mice and other species. In addition, this method has been broadly used as a versatile tool for many fields of mouse micromanipulation. This chapter describes the "one step method" protocol of nuclear transfer in mouse, which combines injection of a donor cell nucleus and enucleation of MII metaphase in a single manipulation procedure. This protocol describes the isolation and collection of oocytes, treatment of donor cells, visualization of spindle-chromosomal complex, direct injection and enucleation, activation of reconstructed embryos and their in vitro culture and transfer into pseudopregnant mice.

Key words Nuclear transfer, Microinjection, Piezo, Cloned mouse, Oocytes

1 Introduction

The first cloned mouse derived from cumulus cell was produced by Dr. Terihiko Wakayama in 1998 using a novel injection method to perform somatic cell nuclear transfer (SCNT) [1]. As this method was set up in the University of Hawaii, it was named "Honolulu technique." In contrast to the electrofusion nuclear transfer method, until then successfully applied to sheep only [2], the "Honolulu technique" relies on two important improvements: Piezo-driven microinjection of donor nuclei and chemical activation of reconstructed embryo. Based on this achievement, other groups have followed in generating cloned mice from embryonic stem (ES) cells [3, 4], fetal neurons [5], immature Sertoli cells [6], tail tip fibroblast, primordial germ cells [7], iPS cells [8, 9], and even "dead" cells isolated from frozen mice [10]. Even though cloned mice can be produced by electrofusion nowadays [11], this method requires live intact donor cell and additional steps compared to "Honolulu technique." Piezo-driven microinjection technology has become broadly applicable for micromanipulation on mouse,

Nathalie Beaujean et al. (eds.), *Nuclear Reprogramming: Methods and Protocols*, Methods in Molecular Biology, vol. 1222, DOI 10.1007/978-1-4939-1594-1_1, © Springer Science+Business Media New York 2015

such as intracytoplasmic sperm injection (ICSI) [12], production of haploid androgenetic embryos [13], nuclear exchange between oocytes, and the injection of ES cells into blastocysts.

More than a decade after the birth of the first cloned mouse, the success rate of mouse SCNT is still very low compared to natural reproduction. Many groups have attempted to increase the efficiency of generating cloned mice by modulating the chromatin state in early cloned embryos or by improving the technical procedure. Several studies have revealed the differences epigenetic states in mouse SCNT versus naturally conceived embryos. For example, such as the level of H3K9 acetylation is lower in SCNT versus control pre-implantation embryos [14]. Based on this finding, the use of histone deacetylases inhibitor (HDACi), such as trischostatin A (TSA) and *m*-carboxycinnamic acid bishydroxamide (CBHA), or Sriptaid, has emerged to improve the mouse SCNT efficiency [15–17]. Treatment of mouse SCNT embryo with HDACi improve greatly the remodeling of nuclear architecture in SCNT early embryos as well as development to term of mouse SCNT embryos [18]. Aberrant X-chromosome inactivation results in reprogramming errors of X-linked genes in mouse male and female SCNT embryos [19]. By knocking-out *Xist* noncoding RNA, Inoue et al. succeeded in obtaining eightfold to ninefold increase in birth rates of male and female SCNT embryos.

On the other hand, an increase of SCNT efficiency can also be achieved by modifying the cloning technique itself or combining other techniques. By aggregation of genetically identical cloned embryos at the 4-cell stage, blastocysts from clone–clone aggregates contain more cells and display higher rates of fetal and postnatal development than control group [20]. Importantly, differentiated somatic cells can be used to establish NT embryonic stem cell lines by SCNT and these nt-ES cells possess the same potential as normal fertilized ES cell lines [21]. Even terminally differentiated cells such as olfactory sensory neuron can be used as donor cells to generate nt-ES cells, which eventually can be used to generate mice by using the tetraploid complementation assay [22, 23].

In 2003, Dr. Qi Zhou combined the injection of the donor nucleus with the enucleation of MII metaphase in oocytes. This modified technique was named the "one step method" and was successfully applied in mouse [4], rat [24] and bovine cloning [25]. In this method, embryos are reconstructed by applying a single micromanipulation step instead of two, used in the Honolulu procedure.

In this chapter, we describe the "one step method" for SCNT in mouse.

To get good and reliable results, the nuclear transfer procedure requires a considerable investment of time and resources for practicing this peculiar skill. Each step of this protocol needs to be carefully checked and tested. As a control, parthenogenetic embryos produced in parallel are necessary.

2 Materials

2.1 Mice Used for Nuclear Transfer

1. B6D2F1(C57BL/6×DBA/2)orB6CBF1(C57BL/6×CBA/J) F1 hybrid mice, between 8 and 12 weeks old, are optimal to be used for obtaining recipient oocytes (*see* **Note 1**). Nuclear transferred embryos derived from these mice oocytes have better development in vitro and in vivo than oocytes from inbreed and outbred mouse strains.

2. ICR (CD-1) mice are usually used as pseudopregnant surrogate mother and foster mothers (*see* **Note 2**).

3. Vasectomized male mouse, from C57BL/6, B6D2F1, or ICR strains.

2.2 Donor Cells

Cumulus derived from Cumulus–Oocytes Complexes (COC) must be used freshly. ES cells and other cultured cells need special treatment. The detailed procedures are explained below.

2.3 Media for Embryo Culture and Manipulation

All media or chemicals used should be embryo or at least cell culture tested grade.

Dissolve the reagents of M2, M16, and KSOM media in ultrapure water (Sigma, W4508) and adjust the pH to 7.2–7.4. All the media can be stored at 4 °C for 2 weeks (*see* **Note 3**). Several companies also provide liquid media.

1. M2: In vitro manipulation and micromanipulation medium.

Components	Final concentration
NaCl	95 mM
KCl	4.8 mM
KH_2PO_4	1.2 mM
$MgSO_4 \cdot 7H_2O$	1.2 mM
Sodium Lactate	23 mM
Sodium Pyruvate	0.3 mM
Glucose	5.6 mM
$NaHCO_3$	4.15 mM
$CaCl_2 \cdot 2H_2O$	1.7 mM
Hepes	20.9 mM
BSA	4 mg/ml

2. M16: culture medium for 1-cell and 2-cell nuclear transfer embryo.

Components	Final concentration
NaCl	95.0 mM
KCl	4.8 mM
KH_2PO_4	1.2 mM
$MgSO_4 \cdot 7H_2O$	1.2 mM
Sodium Lactate	23.0 mM
Sodium Pyruvate	0.6 mM
Glucose	5.6 mM
$NaHCO_3$	25.0 mM
$CaCl_2 \cdot 2H_2O$	1.7 mM
Penicillin G	100.0 IU/ml
Streptomycin Sulfate	0.5 g/ml
BSA	4.0 mg/ml

3. Kalium simplex optimized medium (KSOM)+amino acids (AA): cultural media for cloned mouse embryos from 4-cell until blastocyst stage.

Components	Final concentration
NaCl	95.0 mM
KCl	2.50 mM
KH_2PO_4	0.35 mM
$MgSO_4 \cdot 7H_2O$	0.20 mM
Sodium Lactate	10.0 mM
Sodium Pyruvate	0.3 mM
Glucose	2.8 mM
$NaHCO_3$	25.0 mM
$CaCl_2$	1.71 mM
Glutamine	1.0 mM
EDTA	0.01 mM
Nonessential amino acids	0.5 mM
Essential amino acids	0.5 mM
Penicillin G	100.0 IU/ml
Streptomycin Sulfate	0.5 g/ml
BSA	5.0 mg/ml

4. Activation medium ($CZB + SrCl_2$).

Components	Final concentration
NaCl	82.0 mM
KCl	4.8 mM
KH_2PO_4	1.2 mM
$MgSO_4 \cdot 7H_2O$	1.2 mM
Sodium Lactate	30.1 mM
Sodium Pyruvate	0.26 mM
$NaHCO_3$	25.0 mM
Glutamine	1.0 mM
Penicillin G	100.0 IU/ml
Streptomycin Sulfate	0.5 g/ml
BSA	5.0 mg/ml
$SrCl_2 \cdot 6H_2O$	10.0 mM

See **Note 4**

2.4   Chemicals

1. Hyaluronidase
   Dissolve hyaluronidase in M2 medium directly at 300 IU/ml working concentration. Store it in suitable aliquots at –20 °C and thaw freshly before use. The thawed aliquots can be kept up to 48 h at 4 °C.

2. Cytochalasin B
   Dissolve in DMSO (cell-culture tested) to prepare the stock solution (1 mg/ml) and stored at –20 °C in aliquots (*see* **Note 5**). Before use dilute the stock at 5 μg/ml concentration in M2 and in CZB-activation medium (not for the NT embryo reconstructed with metaphase donor cell, *see* **Note 6**).

3. HDACi
   Dissolve histone deacetylase inhibitors in DMSO, trichostatin A (TSA) at 100 mM, and *m*-carboxycinnamic acid bishydroxamide (CBHA) at 1 μM. Store aliquots at –20 °C. Dilute the compounds freshly 100× in activation and culture media before use.
   Dissolve Scriptaid (Sigma) in DMSO at 25 mM (stock 1). Prepare stock 2 at 250 μm in DMSO. Stock 1 is kept in aliquots at –80 °C and stock 2 at –20 °C for several months. Freshly dilute stock 2 at 1/1,000 in culture and activation medium to a 250 nM final concentration.

4. Demecolcin
   Demecolcin is a microtubule depolymerizing agent which can block donor cells at metaphase stage. Dissolve demecolcin in

DMSO at 10 µg/ml. Store aliquots at –20 °C. Freshly dilute them 200× in cell culture media before use.

5. Mineral Oil
   Light mineral oil is used to cover culture or manipulation medium to prevent evaporation of medium. The oil should be embryo tested grade.

6. Hormones
   Pregnant mare serum gonadotropin (PMSG) and human chorionic gonadotropin (HCG) are dissolved in saline at final concentration of 100 IU/ml and stored at –20 °C.

7. Fluorinert FC-770
   To use piezo-drive pipets, the tip of injection pipet is filled with FC-770 (Sigma) with a density of 1.7–1.8 g/ml.

2.5 Equipment and Tools

1. Inverted microscope
   To visualize meiotic spindle of mouse MII oocytes, inverted microscope should be equipped with differential interference contrast (DIC) optical components, Hoffman modulation contrast (HMC) optics or polarized light system (spindle view microscopy) (Table 1) (*see* **Note 7**).

2. Micromanipulator set
   Coarse manipulator (MMO-202, Narishige) equipped with mot-drive system (MM-89, Narishige).

Table 1
Comparison of different imaging system

Imaging system	Principle	Advantage	Disadvantage
DIC	DIC can recognize difference in refractive index between spindle and ooplasm by producing and shearing plane-polarized light to enhance the contrast in transparent oocytes.	High resolution Noninvasive for live sample	Expensive The oocyte should be transparent Require glass-bottom dish only
HMC	HMC can visualize spindle zone as a 3D shadow by converting optical phase gradients of transparent oocyte into variations of light intensity.	Less expensive Suitable for plastic-bottom and glass-bottom dish	Less sensitive
Polarized light	Polarized light microscopy can visualize bright meiotic spindle filaments on a dark background by occurrence of birefringence (doubly refracting) on spindle.	High degree of sensitive Acquisition of live image for qualitative and quantitative analysis	Expensive CB, low temperature and aging of oocytes can reduce the sensitive of detecting

3. Manual microinjector
   The manual microinjector CellTram Air (Eppendorf) is used for oocytes holding pipet and CellTram Oil (Eppendorf) is used for donor nucleus injection pipet.

4. Piezo device
   Piezo impact drive systems, PMM (Prime Tech Ltd, Ibaraki), and Piezo Xpert (Eppendorf) can both provide good results for zona piercing and oolemma penetration.

5. Stereomicroscope equipped with 0.63× objective.

6. Incubator
   Set humidified CO_2 incubator to 37 °C and 5 % CO_2.

7. Pipet puller P-97 (Sutter Instruments Co.) and microforge MF-900 (Narishige).

8. Glass capillary with inner diameter of 100 μm.

9. Aspirator tube assemblies for microcapillary pipets or pipet holder.

3 Methods

3.1 Superovulation and Collection of Oocytes

1. In standard light–dark cycles in the mouse room, inject 8–10 IU PMSG at 5 ~ 6 PM and 8–10 IU HCG 46–48 h later.

2. 13–15 h after HCG injection, sacrifice the female mice by cervical dislocation and collect oviducts. Retrieve oocyte-cumulus cell complexes (COC) from the oviduct ampulla under stereomicroscope in M2 medium by tearing the ampulla.

3. Place COC in pre-warmed M2 + hyaluronidase (0.3 mg/ml) medium at 37 °C for few minutes to digest the intercellular matrix between oocytes and cumulus cells. Wash oocytes by gently pipetting and transferring them in different separated M2 drops at least three times (*see* **Note 8**). Collect good quality oocytes to M16 medium and recover them in CO_2 incubator for 30 min. This whole procedure is optimally manipulated on a warming stage at 37 °C. Around 10 % of oocytes are abnormal and should be eliminated, such as oocytes with a small cytoplasm, a dark cytoplasm, or an enlarged zona.

3.2 Donor Cell Preparation

3.2.1 Cumulus Cells

1. Collect cumulus cells from the previous step by digesting COC.

2. Resuspended cumulus cells by pipetting and wash them in M2 medium three times to remove residual hyaluronidase. Place a concentrated suspension of cells into the micromanipulation chamber. Cumulus cells can be kept a few hours at 4 °C before use.

3.2.2 Metaphase ES
Donor Cells Preparations

1. Plate $1–2 \times 10^6$ ES cells on a 10 cm tissue culture dish 2 days before nuclear transfer.

2. Change ES cell medium on next day. Confirm ES cells are 70–80 % confluent under inverted microscopy on day 3 (*see* **Note 9**).

3. Synchronized ES cells in metaphase before nuclear transfer. Replace the culture medium of ES cells by fresh pre-warmed culture medium containing 0.05 µg/ml demecolcin. Culture ES cells in demecolcin for 2 h.

4. Collect ES cells in metaphase by gently tapping the sides of the plate. Collect supernatant and rinse twice with ES culture medium.

5. Resuspend the pellet in about 100 µl of M2 and keep the cells at 4 °C (*see* **Note 10**).

6. Add 1–2 µl of cell suspension into micromanipulation chamber each time before nuclear transfer and add fresh suspension aliquot manipulation (*see* **Note 11**).

3.2.3 Fibroblasts or
Other Donor Cells
in Interphase

1. In interphase, cells could be collected after trypsinization. Aspirate cell medium and wash cells with Ca^{2+} and Mg^{2+} free PBS three times. Add 0.5–1.0 ml PBS containing 0.25 % trypsin and 1 mM EDTA into cell culture dish and incubate at 37 °C for several minutes.

2. Use inverted microscopy with phase contrast optics to confirm that 80 % cells form single suspended cells.

3. Add 3 ml culture medium to neutralize the trypsin. Collect cells by aspirating the culture medium into centrifuge tube. Spin down trypsinized cells by centrifugation at $1,000 \times g$ for 5 min. Remove residual trypsin by washing cells in PBS. Resuspend the ES cells with 100 µl culture medium and keep these high concentrated cells at 4 °C.

3.3 Setting
Up Micromanipulation
System

1. Preparation holding and microinjection pipets
   Pull the glass capillaries on a micropipet puller and produce blunt-ended needles on a microforge.
   The outside diameter (OD) and inside diameter (ID) of holding pipet is 80 µm and 20 µm respectively. For injection pipet, the ID is a little smaller than donor cells (e.g., 8 µm for cumulus cells, 12 µm for ES cells and 15 µm for fibroblast cells). Bend both holding and microinjection pipets with an angle of 20°.

2. Assemble micromanipulator chamber and place empty chamber on the stage of inverted microscopy.

3. Back load 2 µl Fc770 into injection pipet and fit the pipe into the actuator of the Piezo system connected with the pipet

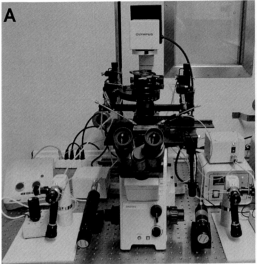

Fig. 1 (**a**) Inverted microscopy with an Eppendorf PiezoXpert unit. (**b**) Injection and holding micropipets in micromanipulation chamber

holder of the CellTram oil microinjector. Push Fc770 to the tip of the pipet. Fit the holding pipet into pipet holder of CellTram air microinjector (Fig. 1a).

4. Put the tips of both holding and injection pipets down to the glass of the micromanipulation chamber. The position of the two tips should be adjusted so as to be at the same Z- and Y-level under the view of inverted microscopy (Fig. 1b).

5. Add 1 ml M2-CB and mineral oil in the micromanipulation chamber. Expel excess air from holding and injection pipets.

3.4 Piezo-Operated Nuclear Transfer

1. Place 10^3–10^4 suspension donor cells in the micromanipulation chamber.

2. Place 20–40 oocytes in the M2 + CB medium in the north sector of micromanipulation chamber. Each group of oocytes should be processed within less than 20 min in vitro.

3. Wait 5 min until CB breaking cytoskeleton of oocytes. Pick single oocytes and rotate it gently by aspirating and blowing with holding pipets from left hand. The metaphase II (MII) spindle can be recognized by DIC or Hoffman optics. Stop rotating and immobilize oocyte until the MII spindle is located at 2 or 4 o'clock (Fig. 2a) (*see* **Note 12**).

4. Approach the zona pellucida with injection pipet from the right-hand side. Push the Fc770 near to the tip of the injection pipet by turning the knob of the manual injector. Drill the zona pellucida by applying a few strong piezo pulses (Fig. 2a) until the zona is pierced (*see* **Note 13**).

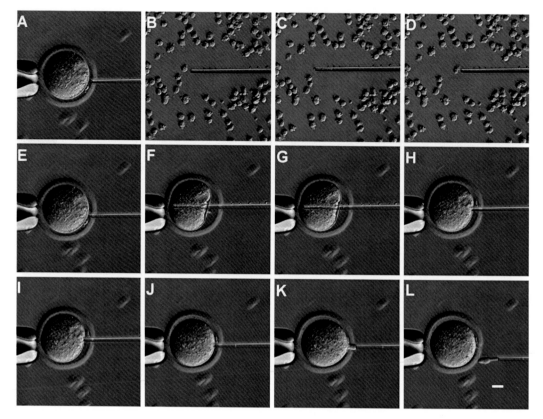

Fig. 2 Steps of somatic cells nuclear transfer in mouse oocytes. Scale bar presents 20 µm

5. Select single donor cell on the bottom of manipulation chamber (Fig. 2b) (*see* **Note 14**). Break the cell membrane and isolate the nucleus from cytoplasm by gently aspirating it in and out of the injection needle once or twice (Fig. 2c, d). Stabilize the donor nucleus inside of injection needle to a distance of 10 µm from the tip.

6. Adjust the focus of invert microscopy and push the injection pipet through the pre-drilled hole of zona pellucida without using any piezo pulse (Fig. 2e) (*see* **Note 15**). Pass injection needle through the oocyte to the opposite side without breaking the ooplasm (Fig. 2f).

7. Puncture the ooplasm by applying a single light piezo pulse. Immediately eject the donor nucleus into the ooplasm (Fig. 2g).

8. Withdraw the injection needle gently until reaching the entry site on the ooplasm (Fig. 2h). Aspirate the ooplasm slightly to seal the hole in oocyte membrane (Fig. 2i) (*see* **Note 16**).

9. Move the injection needle adjacent to the MII spindle and grip the MII spindle by aspirating it into injection needle (Fig. 2j).

Withdraw the injection needle fast but gently and pull MII spindle out of oocytes (Fig. 2k, l). Meanwhile the oocyte membrane can be pinched off.

10. Release the reconstructed oocytes from the holding pipet. Select the next oocyte and repeat the nuclear transfer procedure (*see* **Note 17**).

11. Return the reconstituted oocytes into M16 medium in the incubator and let them recover at least 15 min for embryos reconstructed with metaphase donor cell and 1 h for those reconstructed with interphase cells.

3.5 Activation and Culture of Nuclear Transfer Embryos

1. Prepare 500 μl CZB-SrCl$_2$ activation medium in 4-well plates without oil overlay. Add 5 μg/ml CB (not for the reconstructed oocyte derived from metaphase donor cell) and HDACi (not for ESNT oocytes) into activation medium and equilibrate it for 1 h in incubator at 37 °C under 5 % CO$_2$.

2. Transfer reconstructed embryos into activation medium and culture for 5 h (interphase donor cells or cumulus) or 3 h (metaphase donor cells) in the incubator.

3. Wash the activated reconstructed embryos in M16 droplets more than three times to remove CB. Continue culture reconstructed embryos in M16-HDACi droplets for additional 4 h (*see* **Note 18**).

4. Wash reconstructed embryos in M16 culture medium three times and culture them in M16 until late 2-cell stage.

5. Change the culture medium from M16 to KSOM at late 2-cell stage of reconstructed embryos. Culture reconstructed embryos in KSOM in incubator until blastocyst stage.

3.6 Embryo Transfer

1. Reconstructed embryos can be surgically transferred to surrogate female mice at 1/2-cell stage, 4/8-cell stage, and blastocyst stage depending on the aim of experiment. We routinely transfer 2-cell embryos (24 h post activation) into oviduct ampulla of 0.5 days post-coitum (dpc) pseudopregnant ICR female mice or blastocyst (72–96 h post activation) into uterus of 2.5 dpc pseudopregnant female (*see* **Note 19**).

2. Euthanize surrogate mothers at 18.5 dpc and carefully remove pups from the uterus (*see* **Note 20**). Stimulate the pups to breath by wiping their mouths and noses with soft napkin. Measure the weight of the pup and placenta.

3. Prepare the foster mothers that naturally deliver pups in the same day of the dissection of surrogate mothers. Mix the cloned pups with the litter and bedding material from the cage of foster mother to allow cloned pups to take on the same odor of the other pups (*see* **Note 21**).

4 Notes

1. This strain can produce oocytes efficiently and their oocytes are relatively easy to micromanipulate.

2. ICR female mice have a more developed maternal instinct. They easily accept and breed the cloned pups.

3. Each homemade or fresh commercial medium should be first tested with parthenogenetic embryos. Medium can be used only if 80 % of parthenogenetic embryos develop into blastocysts.

4. For M2, M16 and activation medium, antibiotics are optional.

5. As an inhibitor of actin polymerization, cytochalasin B (CB) makes mouse MII oocytes more flexible.

6. Treatment of reconstructed oocytes with CB in activation medium prevents the exclusion of the second polar body.

7. Optional: a warming stage can be used as it decreases the risk of thermic variations. For enucleation and reconstruction 30 °C is a good compromise to clearly see the spindle, while maintaining the cells in metaphase and preventing oocyte lysis during enucleation/reconstruction.

8. Prepare several 50 μl micro-drops M16 in a 35 mm petri dish and cover it with mineral oil to avoid evaporation. Pre-warm and equilibrate M16 medium (also for KSOM) in CO_2 incubator for a few hours before use.

9. For the best ESNT efficiency, ES cells should be used at early passage (<passage 20).

10. Low temperature slows down the resumption of cell cycle of ES cells, but ES cells cannot be kept at 4 °C for long time. We recommend performing nuclear transfer immediately after resuspending ES cells.

11. This preparation procedure can also be used for iPS cells.

12. The outside border of oocyte ooplasm should be sharply focused through optic while rotating and immobilizing oocyte.

13. The parameters of the piezo to drill the zona pellucida and the cytoplasmic membrane depend of your machine and your injection pipet. The level of vibrations need to be adapted and should remain as low as possible to limit possible deleterious effects of vibrations.

14. Keep the oocyte in the same position with the holding pipet while selecting donor nucleus.

15. Any strong piezo pulse could impair the donor nucleus in the injection needle.

16. Try to aspirate the minimum volume of cytoplasm when sealing the membrane.

17. A skilled operator can finish the whole nuclear transfer procedure in 20–30 s. It is also possible to manipulate in two steps: first enucleating the oocyte and then injecting the donor cell.

18. Pseudo-pronuclei in NT embryos are clearly visible under inverted microscope after activation. Record the number of embryos with pseudo-pronuclei as the number of activated NT embryos.

19. Cloned blastocyst can also be transferred into the oviduct ampulla of pseudopregnant female at 0.5 dpc in case no females mated with vasectomized male 2.5 days in advance are available. Embryo transfer has to be carried out following requirements from local animal care and use committee.

20. Estimate the pregnant progression by gestational age of pseudopregnant female.

21. Remove some pups from foster female's pups if there are many pups in the litter.

Acknowledgements

Z.L. gratefully acknowledges financial support from the Novartis Research Foundation and the Swiss National Science Foundation (National Research Program NRP63—Stem Cells and Regenerative Medicine) that is provided to the laboratory of Prof. Antoine H.F.M. Peters at the FMI. The authors thank Prof. Antoine. H.F.M. Peters for comments on the manuscript.

References

1. Wakayama T, Perry AC, Zuccotti M, Johnson KR, Yanagimachi R (1998) Full-term development of mice from enucleated oocytes injected with cumulus cell nuclei. (Translated from eng). Nature 394(6691):369–374

2. Wilmut I, Schnieke AE, McWhir J, Kind AJ, Campbell KH (1997) Viable offspring derived from fetal and adult mammalian cells. (Translated from eng). Nature 385(6619):810–813

3. Wakayama T, Rodriguez I, Perry AC, Yanagimachi R, Mombaerts P (1999) Mice cloned from embryonic stem cells. (Translated from eng). Proc Natl Acad Sci U S A 96(26): 14984–14989

4. Zhou Q, Jouneau A, Brochard V, Adenot P, Renard JP (2001) Developmental potential of mouse embryos reconstructed from metaphase embryonic stem cell nuclei. (Translated from eng). Biol Reprod 65(2):412–419

5. Yamazaki Y et al (2001) Assessment of the developmental totipotency of neural cells in the cerebral cortex of mouse embryo by nuclear transfer. (Translated from eng). Proc Natl Acad Sci U S A 98(24): 14022–14026

6. Ogura A et al (2000) Production of male cloned mice from fresh, cultured, and cryopreserved immature Sertoli cells. (Translated from eng). Biol Reprod 62(6):1579–1584

7. Miki H et al (2005) Birth of mice produced by germ cell nuclear transfer. (Translated from eng). Genesis 41(2):81–86

8. Liu Z et al (2012) Induced pluripotent stem-induced cells show better constitutive heterochromatin remodeling and developmental potential after nuclear transfer than their parental cells. (Translated from eng). Stem Cells Dev 21(16):3001–3009

9. Zhou S et al (2010) Successful generation of cloned mice using nuclear transfer from induced pluripotent stem cells. (Translated from eng). Cell Res 20(7):850–853

10. Wakayama S et al (2008) Production of healthy cloned mice from bodies frozen at –20 degrees C for 16 years. (Translated from eng). Proc Natl Acad Sci U S A 105(45):17318–17322

11. Yu Y et al (2007) Piezo-assisted nuclear transfer affects cloning efficiency and may cause apoptosis. (Translated from eng). Reproduction 133(5):947–954

12. Yoshida N, Perry AC (2007) Piezo-actuated mouse intracytoplasmic sperm injection (ICSI). (Translated from eng). Nat Protoc 2(2):296–304

13. Li W et al (2012) Androgenetic haploid embryonic stem cells produce live transgenic mice. (Translated from eng). Nature 490(7420):407–411

14. Wang F, Kou Z, Zhang Y, Gao S (2007) Dynamic reprogramming of histone acetylation and methylation in the first cell cycle of cloned mouse embryos. (Translated from eng). Biol Reprod 77(6):1007–1016

15. Kishigami S et al (2006) Significant improvement of mouse cloning technique by treatment with trichostatin A after somatic nuclear transfer. (Translated from eng). Biochem Biophys Res Commun 340(1):183–189

16. Dai X et al (2010) Somatic nucleus reprogramming is significantly improved by m-carboxycinnamic acid bishydroxamide, a histone deacetylase inhibitor. (Translated from eng). J Biol Chem 285(40):31002–31010

17. Van Thuan N et al (2009) The histone deacetylase inhibitor scriptaid enhances nascent mRNA production and rescues full-term development in cloned inbred mice. (Translated from eng). Reproduction 138(2):309–317

18. Maalouf WE et al (2009) Trichostatin A treatment of cloned mouse embryos improves constitutive heterochromatin remodeling as well as developmental potential to term. (Translated from eng). BMC Dev Biol 9:11

19. Inoue K et al (2010) Impeding Xist expression from the active X chromosome improves mouse somatic cell nuclear transfer. (Translated from eng). Science 330(6003):496–499

20. Boiani M, Eckardt S, Leu NA, Scholer HR, McLaughlin KJ (2003) Pluripotency deficit in clones overcome by clone-clone aggregation: epigenetic complementation? (Translated from eng). EMBO J 22(19):5304–5312

21. Wakayama S et al (2006) Equivalency of nuclear transfer-derived embryonic stem cells to those derived from fertilized mouse blastocysts. (Translated from eng). Stem Cells 24(9):2023–2033

22. Eggan K et al (2004) Mice cloned from olfactory sensory neurons. (Translated from eng). Nature 428(6978):44–49

23. Li J, Ishii T, Feinstein P, Mombaerts P (2004) Odorant receptor gene choice is reset by nuclear transfer from mouse olfactory sensory neurons. (Translated from eng). Nature 428(6981):393–399

24. Zhou Q et al (2003) Generation of fertile cloned rats by regulating oocyte activation. (Translated from eng). Science 302(5648):1179

25. Heyman Y et al (2002) Novel approaches and hurdles to somatic cloning in cattle. (Translated from eng). Cloning Stem Cells 4(1):47–55

Chapter 2

Nuclear Transfer in Rabbit

Nathalie Daniel and Patrick Chesné

Abstract

Since 2002, our INRA laboratory (Biologie du Développement et de la Reproduction) has developed a method to produce live somatic clones in rabbit, one of the mammalian species considered as difficult to clone. This chapter presents the technical protocol used nowadays to achieve the goal to obtain cloned embryos able to develop to term using fresh somatic cumulus cells.

Key words Nuclear transfer, Rabbit, Oocytes, Electrofusion, Cumulus

1 Introduction

Reprogramming of somatic cells to pluripotency could be achieved by nuclear transfer (NT) into enucleated oocytes. Compared to the development of natural fertilized embryo, a somatic cell nuclear transfer (SCNT)-derived embryo has the challenge of silencing its somatic-specific genes while reactivating all the embryonic-specific genes. So SCNT is an interesting tool to better understand the signalling pathways involved in nuclear reprogramming and gene regulation in preimplantation embryos.

In this context, rabbit is an attractive model. First the embryonic genome activation (EGA) occurs several cell cycles after fertilization or parthenogenetic activation (8 cell stage), thus providing a good access in time during which processes related to the gradual activation of genes can be studied [1].

Second, when a differentiated cultured somatic cell is used as a donor cell for NT, rabbit is relatively a more difficult species to clone, in comparison to other mammals. Paradoxically, until the end of the 1980s rabbit was the pioneer species in the NT technologies using embryonic cell nuclei [2–6]. Difficulties encountered in the case of SCNT in Rabbit might be useful to better identify or go into details of the critical factors implicated.

Nathalie Beaujean et al. (eds.), *Nuclear Reprogramming: Methods and Protocols*, Methods in Molecular Biology, vol. 1222, DOI 10.1007/978-1-4939-1594-1_2, © Springer Science+Business Media New York 2015

Third, these regulated or unregulated events could be quickly correlated with full-term development, because the gestation is relatively short compared to other species that have EGA around 8 cell stage (such as bovine).

This chapter describes a technical approach to create cloned rabbit embryos using subzonal injection of cumulus cells associated with a cell electrofusion and chemical activation. This protocol of SCNT in rabbit which allows generation of viable and fertile offspring with cumulus as donor cells is inspired from refs. 7, 8.

2 Materials

2.1 Superovulation

1. Animals: Sexually mature New Zealand 1077 INRA rabbit approximately 4.0 kg and more than 6 months old.
2. Follicle Stimulating Hormone (FSH).
3. Buserelin.

2.2 Oocyte Collection

1. Dulbecco's-Phosphate Buffer Saline 1× (D-PBS).
2. Fetal calf serum (FCS).
3. Hyaluronidase (Stock solution 0.5 %): Dissolve 100 mg powder into 20 ml of M199 Hepes. Aliquots are stored at −20 °C.
4. M199.
5. M199 Hepes.

2.3 Embryo Activation and Culture

1. Mineral oil.
2. M199.
3. M199 Hepes.
4. Foetal calf serum (FCS).
5. Hoechst 33342 (Stock solution): Add 1 mg of Hoechst 33342/ml of physiologic serum. Aliquots are stored at +4 °C.
6. Cytochalasin B (Stock solution): Dissolve 1 mg CB into 1 ml DMSO. Aliquots are stored at −20 °C (*see* **Note 1**).
7. DMAP (Stock solution): Dissolve powder to obtain 200 mM in M199. Aliquots are stored at −20 °C (*see* **Note 2**).
8. Cycloheximide (Stock solution): Dissolve 1 mg into 1 ml distilled water. Aliquots are stored at −20 °C (*see* **Note 3**).
9. 0.3 M Mannitol: Dissolve 27.33 g of Mannitol into 500 ml distilled water + 100 μM Mg^{2+} ($MgSO_4$) + 100 μM Ca^{2+} ($CaCl_2$) (*see* **Note 4**).

2.4 Equipment for Micromanipulation Setup

1. Inverted microscope with differential interference contrast (DIC) ×20, ×40 objectives, equipped with fluorescence is recommended. Olympus (example: Model IX-71) or any microscope commonly equipped with DIC.

2. Micromanipulators (left and right) for holding and suction/injection system: Holding system: Manual (Narishige) and Eppendorf pump system (cell tram air); Suction/injection system: Manual (Narishige) and Eppendorf pump system (cell tram oil); Eppendorf pump systems is reliable with Narishige pump systems, depends on technician preferences.

3. Micromanipulation chamber: Petri dish glass bottom.

2.5 Equipment and Products for Oocyte Recovery, Embryo Activation and Culture

1. Incubator at 38 °C, 5 % CO_2.

2. Hot plate.

3. Two stereomicroscopes.

4. Fusion chamber: we use BTX microslide 0.5 mm fusion chamber, model 450.

5. Fusion machine: we use BTX stimulator, model 200.

6. Petri dish Easy-Grip 35 mm.

7. Four-well dishes (MW4).

8. Thermostatic box.

9. Capillaries made of borosilicate glass.

3 Methods

3.1 Superovulation

1. Move females at least 1 week before ovulation from 8 to 16 h light per day regime to initiate follicular growth.

2. Superovulate a female by five subcutaneous injections of FSH: two 5 μg injections at 12 h intervals the first day; two 10 μg injections at 12 h intervals the second day, and finally one 5 μg injection the third day.

3. 10 h later the last FSH injection, inject the female intramuscularly with 0.4 ml of Buserelin to induce ovulation.

4. Euthanize the superovulated female, 15–16 h post-Buserelin injection.

3.2 Micromanipulation Capillaries

1. Pull a borosilicate-glass capillary tubing on a mechanical pipette puller to make holding pipettes. Shorten holding pipette to an outer diameter of 100 μm. Melt the tip using a microforge to obtain an inner diameter of 10–20 μm (*see* **Note 5**).

2. Pull the injection pipette the same way as holding pipette. Shorten suction/injection pipette to an outer diameter of 15 μm, bevel and sharpen the pipette tip with, respectively, a grinder and the microforge (*see* **Note 6**).

3. Make small-bore pipette manually using a Bunsen burner (*see* **Note 7**).

3.3 Set Up the Micromanipulator

1. Micromanipulation setup:
   - Mount the holding pipette in the metal pipette holder of the micromanipulator.
   - Mount the suction/injection pipette into the metal pipette holder of the micromanipulator; fill the pipette with oil contained in the micromanipulator.

2. Prepare the micromanipulation chamber by adding a drop of 1 ml of M199 Hepes + 10 % FCS supplemented with 0.5 µg/ml Hoechst 33342 and 7 µg/ml of cytochalasin B.

3. Insert the holding pipette into the drop and adjust it such that it lies horizontally on the bottom of the micromanipulation chamber.

4. Insert the suction/injection pipette into the drop and adjust it such that it lies horizontally on the bottom of the micromanipulation chamber, parallel to the holding pipette.

5. Cover with mineral oil.

3.4 Oocytes Collection

Prewarm all media to 38 °C.

1. Prepare a MW4 with 0.5 % of hyaluronidase in the first well; the other wells contain M199 Hepes + 10 % FCS. Store until use at 38 °C.

2. Prepare four MW4 containing 0.5 ml of M199 + 10 % FCS per well: Later referred to as MW4 n°1 to n°4. The second well of the MW4-n°1 contains 0.5 µg/ml of Hoechst 33342. Store until use at 38 °C and 5 % CO_2.

3. Take the entire genital tract from the female from the autopsy room (Fig. 1).

4. Move to the laboratory with the genital tract carried in a thermostatic box (38 °C) for oocytes collection.

5. Put the genital tract in a sterile surgical drape prewarmed at 38 °C.

6. Dissect each sides of the genital tract separately. Cut across the uterus horn 2 at 3 cm from the uterotubal junction. To get a good overview of the complete length of the oviduct and the beginning of uterine horn, remove with caution adipose tissues surrounding the horn uterus, oviduct, and also connective tissues. Relax "hair-pin" loops by incision of the tissue thereby freeing the two branches of the loop (Fig. 2 and *see* **Note 8**).

7. Fill a 10 ml syringe with D-PBS and introduce blunt end of the needle into the lumen of the right uterus to retrograde flush Complex Oocytes Cumulus (COCs) (from uterus to oviduct).

8. Gently inject 5 ml of D-PBS and collect COCs in a warm petri dish. Use hot plate to keep petri dishes warm. The COCs are

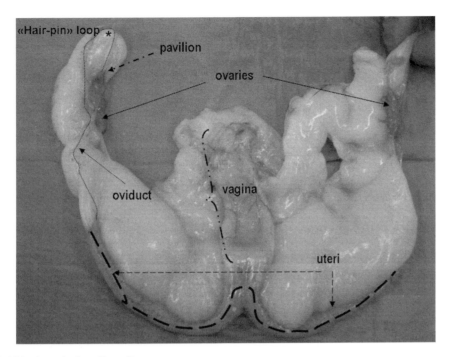

Fig. 1 Rabbit uterus before dissection

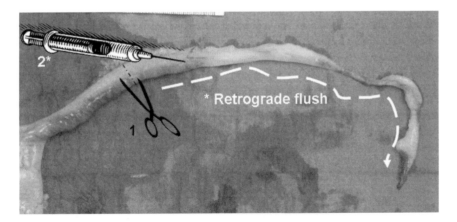

Fig. 2 Dissected oviduct just before retrograde flush

visible without stereomicroscope in the petri Dish (35 mm). The same procedure is then carried out for the left genital tract.

9. Put COCs in the first well containing hyaluronidase 0.5 % and store the MW4 at 38 °C.

10. After few seconds, recover the cumulus and follicular cells (*see* Subheading 3.5) and after 15 min, remove all cumulus cells surrounding oocytes by gentle pipetting with a flame-polished narrow-bore pipette, under stereomicroscope (*see* **Note 9**).

11. Wash the denuded oocytes in M199 Hepes + 10 % FCS.

12. Store the denuded oocytes in M199 + 10 % FCS (MW4 _ n°1) in humidified 5 % CO_2 incubator at 38 °C until use (enucleation).

3.5 Cumulus and Follicular Cells Preparation

1. Use the freshly recovered cumulus cells as a source of nuclei.

2. Incubate recovered COCs in pre-warmed 0.5 % hyaluronidase for a few seconds (*see* Subheading 3.4).

3. Remove follicular and cumulus cells from hyaluronidase as soon as they are detached.

4. Wash them in D-PBS and disperse them carefully in micromanipulation chamber. Keep them away from future oocytes' place (*see* next Subheading 3.6).

3.6 Enucleation

1. Incubate oocytes at 38 °C in M199 + 10 % FCS supplemented with 0.5 μg/ml Hoechst for 10 min.

2. Place a batch of oocytes (num = 20) in the drop of the micromanipulation chamber (*see* **Note 10**).

3. Aspirate one oocyte by the holding pipette and move it to position the Metaphase II plate (MII) or polar body at 3 o'clock.

4. Remove MII (Fig. 3). It is sometimes difficult to visualize the MII, because of the presence of dark cytoplasmic granules. To overcome this problem, perform the enucleation using low ultraviolet light. Detect the MII by brief exposure of the oocyte at UV light. When MII is detected, aspirate it under visible light using the enucleation pipette with minimal volume of oocyte cytoplasm.

5. Visualize the removed condensed chromatin inside the enucleation pipette under UV light, if not distinguishable with classic light transmission.

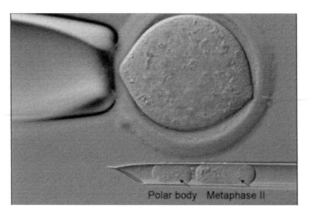

Fig. 3 Rabbit ooplasm after removing of polar body and metaphase II (distinguishable in the enucleation pipette)

6. Remove polar body. When possible, remove the polar body at the same time as MII removing. We talk about ooplasm when oocyte is without polar body and metaphase II (Fig. 3 and *see* **Note 11**).

3.7 Nuclear Transfer and Fusion

1. Aspirate a donor cell (cumulus) with the same pipette used for enucleation. The cell diameter is around 15 μm.

2. Introduce gently the cumulus cell beneath the zona pellucida.

3. Reconstruct all the batch of oocytes in the same way.

4. After reconstruction, keep reconstructed oocytes in M199 + 10 % FCS at 38 °C, 5 % CO_2 during the preparation of fusion materials described below (MW4 _ n°1 except the second well).

5. Prepare a MW4: two wells containing M199 Hepes + 10 % FCS and two wells containing Mannitol solution (Later referred to as MW4 fusion).

6. Prepare the fusion chamber by connected electrodes to the stimulator and cover the electrodes with mannitol solution.

7. Place reconstructed oocytes in MW4 fusion in M199 hepes + 10 % FCS.

8. Rinse twice reconstructed oocytes in MW4 fusion in Mannitol solution.

9. Set the stimulator: 3.2 kV/cm DC pulse 20 μs × 3.

10. Place by batch (number = 5), reconstructed oocytes between the two electrodes of the fusion chamber.

11. Turn each reconstructed oocytes correctly between the two electrodes (Fig. 4).

12. Apply electrofusion.

13. Place the batch of reconstructed oocytes in MW4 fusion in M199 hepes + 10 % FCS waiting for all remaining embryos are electrofused.

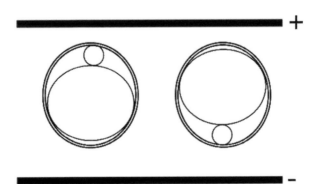

Fig. 4 Correct orientation of reconstructed oocytes

14. Repeat procedure with the remaining batches of reconstructed oocytes.

15. Incubate all reconstructed oocytes in M199 + 10 % FCS (MW4 _ n°2) for 1 h at 38 °C, 5 % CO_2 (*see* **Note 12**).

3.8 Activation

1. Repeat the same electric procedure on fused "embryos."

2. Incubate fused "embryos" in M199 + 10 % FCS containing DMAP (2 mM) + cycloheximide (5 µg/ml) (MW4 _ n°3) for 1 h at 38 °C, 5 % CO_2 (*see* **Note 13**).

3.9 Embryo Culture

1. After activation, rinse embryos in M199 + 10 % FCS (MW4 _ n°4) and culture them in microdrops (40 µl) of M199 + 10 % FCS under mineral oil at 38 °C and 5 % of CO_2 for 96 h until blastocyst formation (*see* **Note 14**).

4   Notes

1. Wear a Personal Protective Equipment when weighing Cytochalasin, according to manufacturer recommendations. This product is considered as Cancerogenic–Mutagenic–Reprotoxic agent.

2. Wear a Personal Protective Equipment when weighing Cycloheximide, according to manufacturer recommendations. This product is considered as Cancerogenic-Mutagenic-Reprotoxic agent.

3. Concerning 6-DMAP, in suspension it is hard to obtain a good mixing: a continuous stirring at 37 °C in necessary.

4. Mannitol solution is prepared and stored in plastic tubes. Indeed, if the tube is a glass tube, a crystallization of $CaCl_2$ and $MgSO_4$ can take place in contact with the glass.

5. The holding pipette is used to hold gently but firmly the oocyte in the position during the micromanipulation.

6. The suction/injection pipette is used to penetrate the oocyte zona pellucida and ooplasm, to remove the metaphase II, polar body and to introduce donor cell in the subzonal space.

7. The small-bore pipette is used to denude, transport oocytes and cloned embryos in all experiments.

8. Relaxing of hair-pin loops will greatly facilitate the subsequent flushing of the oviduct. All the process of dissection occurs rapidly and under a hot plate to prevent minimal thermal variations.

9. A special care should be taken during the denudation of oocytes. Indeed, oocytes are very sensitive to the ambient temperature; they must be manipulated very quickly with minimal

thermal variations to avoid uncontrolled and spontaneous activation.

10. Oocytes are handled in groups of 20 to reduce the time between enucleation and fusion.

11. The polar body has to be absolutely because polar bodies fuse more easily than cumulus cells and they are subjected to develop to blastocyst stage over 30 % [9].

12. After electrofusion, reconstructed embryos are incubated 1 h in classical culture conditions. During this incubation, the fused nucleus is subjected to a metaphasic ooplasm so it will undergo nuclear envelope breakdown and chromatin condensation due to the high level of MPF activity. This stage is called PCC: Premature Chromatin Condensation.

13. These two drugs allow the passage from metaphasic to interphasic cytoplasmic environment and the remodelling of the nucleus.

14. If an embryo transfer is required, the embryos (20–25 per female) are transferred at 2–4 cell stage in the oviduct via the infundibulum. The recipient female is prepared with a delay of 22 h (Buserelin injection) compared to the preparation of donor oocytes female.

Disclosure Statement

License: INRA (FR)—BIOPROTEIN TECHNOLOGIES (FR). Rabbit nuclear cloning method and uses threrof. EP 1465994. 2002.01.10.

References

1. Pacheco-Trigon S, Henneguet-Antier C et al (2002) Molecular characterization of genomic activities at the onset of zygotic transcription in mammals. Biol Reprod 67(6):1907–1918

2. Stice SL, Robl JM (1988) Nuclear reprogramming in nuclear transplant rabbit embryos. Biol Reprod 39:657–664

3. Collas P, Robl JM (1990) Factors affecting the efficiency of nuclear transplantation in the rabbit embryo. Biol Reprod 43:877–884

4. Collas P, Robl JM (1991) Relationship between nuclear remodeling and development in nuclear transplant rabbit embryos. Biol Reprod 45:455–465

5. Collas P, Fissore R et al (1993) Preparation of nuclear transplant embryos by electroporation. Anal Biochem 208:1–9

6. Yang X, Jiang S et al (1992) Nuclear totipotency of cultured rabbit morulae to support full-term development following nuclear transfer. Biol Reprod 47:636–643

7. Chesné P, Adenot PG, Viglietta C et al (2002) Cloned rabbits produced by nuclear transfer from adult somatic cells. Nat Biotechnol 20(4):366–369

8. Challah-Jacques M, Chesne P, Renard JP (2003) Production of cloned rabbits by somatic nuclear transfer. Cloning Stem Cells 5(4):295–299

9. Chesné P, Chrenek P et al (2006) Cloning in the rabbit: present situation and prospects. In: Inui A (ed) Epigenetic risks of cloning, 1st edn. Taylor and Francis, Boca Raton, FL, pp 71–96

Chapter 3

Nuclear Transfer in Ruminants

Joon-Hee Lee and Walid E. Maalouf

Abstract

Ruminants were the first mammalian species to be cloned successfully by nuclear transplantation. Those experiments were designed to multiply high merit animals (Willadsen, Nature 320(6057):63–65, 1986; Prather et al., Biol Reprod 37(4):859–866, 1987; Wilmut et al., Nature 385(6619):810–813, 1997). Since then, cloning has provided us with a vast amount of knowledge and information on the reprogramming ability of somatic cells to different cell types which became an important basis for stem cell research and human medicine. Nowadays, the goals of most nuclear transfer work vary widely but in most cases the micromanipulation procedures remain the same. However, differences between species require different technical considerations. In this chapter, we describe in detail somatic cell nuclear transfer which is the foremost method for cloning ruminants with specific reference to sheep and cattle.

Key words Nuclear transfer, Ovine, Fibroblast, Fusion

1 Introduction

The cloning of ruminants started by the work Steven Willadsen in sheep on electrofusion of embryonic cells with enucleated metaphase two oocytes, which led to the birth of identical lambs [1]. Similar work by Randall Prather in cattle followed closely [2].

In the 1990s, assisted reproduction, genetic engineering, transgenesis, cell biology, and other related research were progressing. However, the idea of dedifferentiating a somatic cell into an embryonic state was still science fiction. It was the work of Keith HS Campbell and his colleagues [3, 4] that started the revolution and the progress was first marked by the birth of two cloned Welsh mountain ewes—Megan and Morag—from in vitro cultured and differentiated day nine embryo disk cells [4]. Then, the breakthrough came about with the birth of Dolly the sheep from

---

Dedicated to Professor Keith HS Campbell
"Sheep in Scotland are very, very, very cheap."
Keith HS Campbell, NY Times, 1997.

Nathalie Beaujean et al. (eds.), *Nuclear Reprogramming: Methods and Protocols*, Methods in Molecular Biology, vol. 1222, DOI 10.1007/978-1-4939-1594-1_3, © Springer Science+Business Media New York 2015

cultured somatic mammary cells from a 6-year-old pregnant ewe [5]. Cloning technology paved the way to a profound understanding of nuclear remodelling and reprogramming and a more thorough knowledge on the mechanisms of early development.

In this chapter, we describe in detail somatic cell nuclear transfer which is the foremost method for cloning ruminants with specific reference to sheep and cattle.

2   Materials

2.1   Ovary Collection Medium (PBS)

1. 137 mM Sodium Chloride (NaCl), 2.7 mM Potassium Chloride (KCl), 10 mM Sodium Phosphate dibasic (Na_2HPO_4), and 2 mM Potassium Phosphate Monobasic (KH_2PO_4).

2. For 1 l of volume, add 80.0 g of NaCl, 2 g of KCl, 14.4 g of Na_2HPO_4, and 2.4 g of KH_2PO_4, make up to 10.0 l final volume with distilled water including pH adjustment to 7.4 with 5 N Sodium Hydroxide (NaOH, *see* **Note 1**).

3. Sterilize by autoclaving for 20 min at 15 lb/in.^{2}.

2.2   Oocyte Dissection Medium

1. For oocyte dissection medium, add 10 % heat-inactivated fetal calf serum (FCS) to Tissue Culture Media 199 (TCM 199).

2. For 20 ml of volume, add 2.0 ml of fetal calf serum to 18.0 ml of TCM 199.

3. Sterilize by filtration through a 0.22 μm Millipore filter and warm up to a temperature of 39 °C prior to use.

2.3   Oocyte Maturation Medium

1. Sodium pyruvate stock: add 11 mg of sodium pyruvate to 1 ml of TCM 199. The sodium pyruvate stock will be diluted in maturation medium to a working concentration of 0.2 mM.

2. 17β-estradiol stock: 1 mg 17β-estradiol to 1 ml of ethyl alcohol. Stock solution can be stored at –20 °C for a maximum of 1 month, and it is diluted in maturation medium to a working concentration of 1 μg/ml.

3. Gentamycin stock: add 50 mg of gentamycin to 1 ml of normal saline. The stock solution can be stored at –4 °C, and it is diluted in maturation medium to a working concentration of 50 μg/ml.

4. Cysteamine (2-Mercaptoethylamine) stock: add 7.714 mg of cysteamine to 1 ml of TCM 199. The stock solution can be dispensed in 20 μl aliquots, and stored at –20 °C, and it is then diluted in maturation medium to a working concentration of 100 μM.

5. FSH (Follicular stimulating hormone) stock: add 400.0 mg of FSH (Folltropin®-V; Vetrepharm) to 80 ml of normal saline.

The stock solution can be dispensed in 20 µl aliquots, and stored at –20 °C. FSH stock will be diluted in maturation medium to a working concentration of 5 µg/ml.

6. LH (Luteinizing hormone) stock: add 25.0 mg of FSH (Lutropin®-V; Vetrepharm) to 5.0 ml of normal saline. The stock solution can be dispensed in 20 µl aliquots, and stored at –20 °C. LH stock will be diluted in maturation medium to a working concentration of 5 µg/ml.

7. Working oocyte maturation medium: add 1 ml FCS, 10 µl FSH, 10 µl LH, 10 µl 17β-estradiol, 30 µl sodium pyruvate, and 10 µl cysteamine, to 9 ml TCM 199. Sterilize by filtration through a 0.22 µm Millipore filter and equilibrate in an atmosphere of 5 % CO_2 in air with maximum humidity at a temperature of 39 °C for at least 2 h prior to use.

2.4 Composition of Modified Synthetic Oviduct Fluid (mSOF) Medium

Component	MW	Concentration	g/l
NaCl	58.44	107.63 mM	6.2899
KCl	74.55	7.16 mM	0.5338
KH_2PO_4	136.10	1.19 mM	0.1620
$MgSO_4$	120.40	1.51 mM	0.1818
$CaCl_2 \cdot 2H_2O$	147.00	1.78 mM	0.2617
Sodium lactate (60 % syrup)	112.10	5.35 mM	757 µl
$NaHCO_3$	84.01	5.00 mM	0.4200
Na-pyruvate	110.00	7.27 mM	0.7997
L-Glutamine	146.10	0.20 mM	0.0292
Tri-Sodium-citrate	294.10	0.34 mM	0.1000
HEPES	238.30	20.00 mM	4.7660
Phenol-red	–	10.0 µg/ml	10 mg
H_2O	–	–	Up to 1.0 l

Adjust osmolarity to 270–280 mOsm and to pH 7.4 with 1 N sodium Hydroxide.

Store at 4 °C for a maximum of 1 week.

Before use add 45 µl/ml BME amino acids, 5 µl/ml MEM amino acids, and 5 % heat-inactivated FBS.

Sterilize by filtration through a 0.22 µm Millipore filter and equilibrate in an atmosphere of 5 % CO_2 in air with maximum humidity at a temperature of 39 °C for at least 2 h prior to use.

2.5   Composition of HEPES-Buffered Synthetic Oviduct Fluid (H-SOF) Medium

Component	MW	Concentration	g/l
NaCl	58.44	107.63 mM	6.2899
KCl	74.55	7.16 mM	0.5338
KH_2PO_4	136.10	1.19 mM	0.1620
$MgSO_4$	120.40	1.51 mM	0.1818
$CaCl_2 \cdot 2H_2O$	147.00	1.78 mM	0.2617
Sodium lactate (60 % syrup)	112.10	5.35 mM	757 µl
$NaHCO_3$	84.01	5.00 mM	0.4200
Na-pyruvate	110.00	7.27 mM	0.7997
L-Glutamine	146.10	0.20 mM	0.0292
Tri-Sodium-citrate	294.10	0.34 mM	0.1000
HEPES	238.30	20.00 mM	4.7660
Phenol-red	–	10.0 µg/ml	10 mg
H_2O	–	–	Up to 1.0 l

Adjust osmolarity to 270–280 mOsm and to pH 7.4 with 1 N sodium Hydroxide.

Store at 4 °C for maximum 2 weeks.

Add 5 µl/ml 100× MEM amino acids and 3 mg/ml BSA before use.

Sterilize by filtration through a 0.22 µm Millipore filter.

Equilibrate at a temperature of 39 °C prior to use.

2.6   Other Stock Solutions

1. Hyaluronidase stock (300 IU/mg): add 100 mg hyaluronidase and 1.1 g PVP to 110.0 ml Ca^{2+}-, Mg^{2+}-free Dulbecco's phosphate-buffered saline (DPBS). The stock solution can be dispensed in 1 ml aliquots and stored at –20 °C.

2. Cytochalasin B stock: add 10 mg cytochalasin B to 1.0 ml DMSO. The stock solution can be dispensed in 10 µl aliquots, and stored at –20 °C. Cytochalasin B stock will be diluted in H-SOF at a working dilution of 5 µg/ml.

3. Hoechst 33342 stock: add 25 mg Hoechst 33342 to 5 ml ddH$_2$O. The stock solution can be dispensed in 10.0 µl aliquots, and stored at –20 °C. Cytochalasin B stock will be diluted in H-SOF at a working dilution of 5 µg/ml.

4. Ca^{2+} Ionophore (A23187) stock: add 5 mg Ca^{2+} Ionophore (A23187) to 1.91 ml DMSO. The stock solution (5 mM) can be dispensed in 10.0 µl aliquots, and stored at –20 °C. Ca^{2+} Ionophore (A23187) stock will be diluted in H-SOF at a working dilution of 5 µM.

5. Cycloheximide (CHXM) stock: add 5 mg cycloheximide to 1.0 ml DPBS. The stock solution can be dispensed in 20 μl aliquots, and stored at –20 °C. Cycloheximide (CHXM) stock will be diluted in mSOFaaci at a working dilution of 10 μg/ml.

6. Fusion medium (Mannitol): add 300 mg mannitol, 0.03675 mg $CaCl_2·2H_2O$, and 1.1 g $MgSO_4$, to 110 ml ddH_2O. The stock solution can be dispensed in 1 ml aliquots and stored at –20 °C.

7. Washing medium: supplement 0.1 % (v/v) penicillin/streptomycin to Ca^{2+}-, Mg^{2+}-free Dulbecco's phosphate-buffered saline (DPBS).

8. Isolation medium: supplement 0.25 % (v/v) trypsin–EGTA to DPBS.

9. Dulbecco's modified Eagle medium (DMEM): supplement with 1 % β-mercaptoethanol, 2 mM L-glutamine, 1 % (v:v) penicillin/streptomycin to DMEM.

10. Gelatin solution: add 0.05 g of gelatin to 50 ml Milli-Q water. Autoclave for 30 min within 2 h after mixing. Store the 0.1 % gelatin solution in a refrigerator until use.

2.7 Preparation of Holding Pipettes for Micromanipulation

1. Pull borosilicate glass capillaries (1 mm outer diameter (o.d.)×0.58 mm inner diameter (i.d.)×10 cm without inner filament) by hand over a small flame to give 120–150 μm (o.d.).

2. Mark the capillaries and break at the required size, approximately 150 μm (o.d.), using a diamond pencil.

3. Mount the pipette onto the micro forge and heat until the open tip is almost closed, ensuring an internal diameter of approximately 20 μm.

4. Bend the pipette approximately 2 mm from the tip at an angle 30–45°, this allowing the tip to be manipulation chamber when mounted on the microscope.

2.8 Preparation of Enucleation Pipettes for Micromanipulation

The preparation of enucleation (or injection) pipettes is more elaborate than the holding pipettes.

1. Pull glass capillaries (1 mm outer diameter (o.d.)×0.58 mm inner diameter (i.d.)×10 cm without inner filament) using the flaming micropipette puller to give an initial taper, which reduces the diameter of the capillary to slightly greater than the diameter required, with the second taper being almost parallel.

2. Break a drawn capillary at the required size, between 20 and 30 μm using fine forceps.

3. Mount the pipette in the micro grinder at a 50° angle and its broken tip ground at medium speed, ensuring a continuous water flow over the surface where the pipette is being ground.

4. Dip the ground angled pipette tip into hydrofluoric acid (*see* **Note 2**) for 5 s whilst continuously blowing air through

the capillary using a 10 ml syringe (this prevents acid entering the inside of the pipette), this thins the capillary walls.

5. Wash the pipette washed five times with distilled water to remove the acid and dried. Sigmacote (SL-2) is then drawn into the capillary.

6. Return the pipette to the microforge, the tip of the pipette is touched against a heated glass bead and then pull out to a sharp point.

7. Bend the pipette at an angle of 30–45° like the holding pipette.

2.9 Manipulation Chamber

1. Use a 92 × 10 mm petri dish for manipulation chamber.

2. Place 50 μl drops of warmed manipulation medium (39 °C) in the center of dish and cover in warmed (37 °C) mineral oil, to prevent evaporation of the manipulation medium.

3. Place the chamber onto a heated stage on an inverted microscope fitted with hydraulic micromanipulators for oocyte manipulation.

3   Methods

3.1 Isolation of Primary Fetal Fibroblast Cells

1. Establish primary fetal fibroblast cultures from a 30-day-old fetus as previously described [6].

2. Dissect the fetus from the uterus and then transport on ice from the slaughterhouse to the laboratory.

3. Dissect the fetus free of all external membranes, eviscerate, and decapitate for isolation of cells from the carcass.

4. Wash the tissue in Ca^{2+}-, Mg^{2+}-free Dulbecco's phosphate-buffered saline (DPBS) at 37 °C, then with alcohol, and finally with washing medium.

5. Cut the tissue into small pieces using sterilize scissors in and incubate at 37 °C for 5 min.

6. Transfer the suspension into 15 ml centrifugation tubes and allow to sediment.

7. Separate dissociated cells from larger pieces of tissue by centrifugation at $300 \times g$ for 5 min.

8. Wash dissociating tissue in DPBS and place it on ice. The procedure is repeated several times.

9. Wash supernatants several times in DPBS, the final supernatant is centrifuged at $300 \times g$ for 10 min to obtain a cell pellet.

10. Resuspend the pellet in 10 ml DMEM and 10 % FCS, transfer into 75 cm^2 tissue culture flasks and incubate for 24 h at 37 °C in a humidified atmosphere of 5 % CO_2 in air.

11. Replace the culture medium with fresh medium after 24 h.

12. Primary cultures are grown on tissue culture plastic coated with 0.1 % gelatine until the first passage; in subsequent passages the culture of cells only required standard plastic tissue culture.

3.2 Passage and Cryopreservation of Primary Fetal Fibroblasts

1. When confluent fibroblast monolayers are obtained, the cells are washed with 2 ml Ca^{2+}-, Mg^{2+}-free DPBS (39 °C) twice and incubated in the 0.25 % trypsin–EGTA solution at 39 °C for 2–3 min.

2. The solution containing the detached cells is aspirated into a 50 ml conical tube and diluted with 8 ml of DMEM containing 10 % FCS (to inactivate the trypsin). To detach the remaining cells this treatment is repeated three times.

3. The cell suspension is centrifuged at $300 \times g$ for 5 min and the cell pellets are resuspended in 30 ml of fresh culture medium.

4. Cells are passaged into 3×75 cm³ flasks and culture continued.

5. At subconfluency, cells are cryopreserved at passage 1 in aliquots of ~1×10^6 per vial. 1×10^6 harvested cells are resuspended in 450 μl serum/cryotube (1 ml tubes), the tubes are placed at 4 °C for 20 min and then 10 % (50 μl) dimethyl sulfoxide (DMSO) added.

6. The tubes are mixed and immediately stored in polyvinyl bubble wrap at –80 °C for 24 h.

7. The vials are then transferred to liquid N_2 and stored until required.

3.3 Culture of Cells for Use as Nuclear Donors

1. For each experiment, a vial of primary fibroblasts is placed in water bath (39 °C) for thawing and cultured in DMEM supplemented with 1 % β-mercaptoethanol, 2 mM L-glutamine, 1 % (v:v) penicillin/streptomycin, and 10 % FCS.

2. Cells are passaged every 3–4 days by trypsin solution (0.25 %) and plating at split ratio of 1:10.

3. Primary fetal fibroblasts at passages 4–5 are cultured until approximately 80–90 % confluence; quiescence (presumptive G0) is then induced by reducing the concentration of FBS to 0.1 % for a further 2–3 days.

4. Immediately before use as nuclear donors, a single cell suspension is prepared by trypsinization.

5. The cells are and pelleted and resuspended in DMEM plus 0.1 % FBS and remained in this medium at 37 °C until used as nuclear donors.

3.4 Oocyte Recovery

1. Collect ovaries from mature female ruminants at a local slaughterhouse, immediately place them into phosphate-buffered saline (PBS) at 25 °C, and transport to the laboratory within 2–3 h of collection.

2. Trim excess tissue of the ovaries with scissors and wash twice with sterile PBS (37 °C).

3. Aspirate antral follicles (2–3 mm diameter) using a 10 ml syringe and a hypodermic needle (21 gauge, 1.5 mm internal diameter).

4. Transfer aspirated follicular fluid into a 50 ml conical polystyrene tube and allow settling for 10 min.

5. Remove three quarters of the supernatant.

6. Dilute the remaining follicular material with an equal volume of dissection medium (39 °C) and transfer it into a 92 × 10 mm petri dish.

7. Examine the diluted follicular fluid for oocytes under a dissecting microscope. Oocyte quality is assessed on the basis of morphological appearance.

3.5 In Vitro Maturation of Ovine Oocytes

1. Wash selected oocytes three times in dissection medium (20 ml, 39 °C) and then twice in maturation.

2. For maturation, culture groups of 40–45 oocytes in 500 μl of maturation medium overlaid with mineral oil in 4-well dishes at 39 °C in a humidified atmosphere of 5 % CO_2.

3. Sterilize dissection and maturation media by filtration through a 0.22 μm filter, and equilibrate in an atmosphere of 5 % CO_2 in air with maximum humidity at a temperature of 39 °C for at least 2 h prior to use.

4. Select only good quality oocytes surrounded by at least three layers of cumulus cells and exhibiting a homogeneous cytoplasm.

3.6 Oocyte Enucleation

1. At 15 h post onset of maturation (hpm), place oocytes into 400 μl of H-SOF medium containing 300 IU/ml of hyaluronidase (type IV) in a 15 ml conical polystyrene tube, incubate at 39 °C for 2 min and then vortex for 4–5 min.

2. Wash the denuded oocytes in H-SOF medium supplemented with 4 mg/ml BSA and return to fresh maturation medium.

3. At regular intervals, pre-incubate batches of 15–20 oocytes extruding anaphase/telophase I (AI-TI) spindle in H-SOF containing 5 μg/ml bisbenzimide (Hoechst 33342) for 15 min at 39 °C on the heated stage.

4. After incubation, transfer each batch of treated oocytes to manipulation chamber.

5. Using 20× magnification, pick up and attach a single oocyte to the holding pipette using negative pressure.

6. Change the magnification to 40× and focus on the oocyte held by the pipette.

7. Bring the enucleation pipette into focus. Using the enucleation pipette, rotate the oocyte into a position where the extruding AI-TI spindle can be aspirated into the pipette (Fig. 1a).

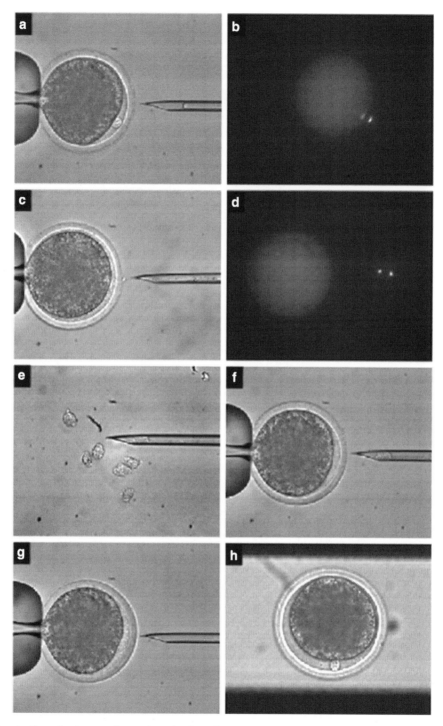

Fig. 1 Enucleation of ovine oocytes and production of oocyte somatic cell couplets. (**a**) Holding the Anaphase 1-Telophase I (AI-TI) oocyte prior to enucleation (the extruding AI-TI spindle). (**b**) Localization of extruding AI-TI spindle by epifluorescence. (**c**) Enucleated oocytes. (**d**) Confirmed by epifluorescnece (the extruding AI-TI spindle in the pipette). (**e**) Somatic cells in suspension. (**f**) Transfer of donor cell into enucleated oocyte. (**g**) Cells after transfer should be in close contact with the oolema. (**h**) The couplet is positioned between the electrodes (*dark lines*) for fusion

8. Insert the enucleation pipette through the zona pellucida at a point opposite the holding pipette.

9. Manipulate the enucleation pipette into a position next to the extruding AI-TI spindle, apply a small amount of negative pressure, and aspirate the AI-TI spindle from directly beneath it into the pipette (Fig. 1c).

10. Withdraw the enucleation pipette from the oocyte and remove the oocyte from the field of view (*see* **Note 3**).

11. Turn off the transmitted light source, change to UV illumination, and examine the aspirated karyoplast (while inside the pipette) for fluorescence using filter block UV-2A. If the metaphase has been removed, it will fluoresce with a blue color; the spindle will also be visible (Fig. 1b, d).

12. Move the enucleated oocyte to the right-hand side of the chamber and discard the aspirated karyoplast from the pipette.

13. Enucleation of oocyte is confirmed by visualization of the DNA in the aspirated karyoplast using a short exposure to UV light (0.1 s).

14. Enucleated oocyte is returned to maturation medium and culture until reconstruction and fusion.

3.7 Production of Oocyte Cell Couplets

1. Prepare a manipulation chamber, as for the enucleation procedure, containing H-SOF 4 mg/ml BSA.

2. Transfer the donor cells into the upper left-hand corner of the chamber and a group of enucleated oocytes into the right hand corner of the chamber.

3. Pick up an encleated oocyte with the holding pipette.

4. Move the holding pipette to the top.

5. Focus the microscope onto the bottom of the chamber.

6. Move the injection pipette to the chamber, maneuver it to the cells, and gently aspirate two to three suitable cells into a 25–30 μm o.d. pipette (Fig. 1e).

7. Refocus on the enucleated oocyte and move the injection pipette until it is in focus.

8. Insert the injection pipette through the hole previously made in the zona pellucida. While holding the pipette against the cytoplasm, inject a donor cell into the perivitelline space of the oocyte (Fig. 1f).

9. The cell is wedged between the zona pellucida and the cytoplast membrane to facilitate close membrane contact for subsequent fusion (Fig. 1g).

10. Upon completion of each batch (15–20 oocytes), wash and then incubate cytoplast-donor couplets in H-SOF without CB for 10 min until the time of fusion.

3.8 Cell Fusion

1. Cell fusion is carried out at room temperature, in a chamber with two platinum electrodes 200 μm apart overlaid with 500 μl of fusion medium.

2. Using a hand-drawn capillary mouth pipette, wash two to three couplets in the fusion medium outside the electrodes and then place them between the electrodes.

3. Align them with an AC pulse (5 V/cm, 5 s) so that the contact surface between the cytoplast and the donor cell is parallel to the electrodes (Fig. 1h, *see* **Note 4**).

4. Quiescent primary fetal fibroblasts used as nuclear donors are electrically fused to enucleated cytoplasts with two DC pulses of 1.25 kV/cm for 30 μs using an Eppendorf Multiporator.

5. After the electrical stimulus, remove the reconstructed embryos from the chamber and gently wash each batch with H-SOF medium.

6. Transfer them into a 50 μl drop of mSOFaaci containing 4 mg/ml FAF-BSA medium until activation.

7. Pulsed couplets are examined 30 min after the fusion pulse for fusion.

8. Unfused couplets are re-pulsed once with the above parameters.

3.9 Chemical Activation

1. At 1 h post fusion, activate reconstructed embryos in H-SOF medium plus 5 μg/ml calcium ionophore (A23187) for 5 min at 39 °C, followed by culture in mSOFaaci medium supplemented with 10 μg/ml of cycloheximide (CHXM) and 7.5 μg/ml cytochalasin B (CB) for 5 h at 39 °C in a humidified atmosphere of 5 % CO_2.

3.10 Culture of Reconstructed Embryos

1. Following activation, transfer the reconstructed embryos into 50 μl drops of mSOFaaci medium containing 4 mg/ml FAF-BSA covered with mineral oil and culture in a humidified atmosphere of 5 % O_2, 5 % CO_2, and 90 % N_2 at 39 °C.

2. On day 2 of culture, cleavage is assessed and 5 % FBS added to the culture medium.

3. On day 7, blastocyst stage embryos are surgically transferred into the uterus of synchronized surrogate ewes (*see* **Note 5**).

4 Notes

1. Dilute solutes in 1 l of distilled water, adjust the pH with 5 N sodium hydroxide, then make up the 10 l final volume.

2. Hydrofluoric acid is corrosive. It can cause severe burns to the skin and eyes. If it comes into contact with skin, you may not feel pain at once. Hydrofluoric acid is also highly irritating to

the respiratory system and very toxic if swallowed. Users should have clear first aid measures in place.

3. Reduce the exposure of enucleated oocytes to the UV light directly to avoid potential long term damage by removing the oocyte from the field of view and observing the enucleated spindle in the aspirated portion inside the enucleation pipette.

4. Precise orientation is necessary for fusion to occur.

5. One to three blastocysts are transferred per recipient.

Acknowledgements

The authors dedicate this book chapter to their late Ph.D. Supervisor, Professor Keith HS Campbell (1954–2012).

This work was partly carried out with the support of "Cooperative Research Program for Agriculture Science & Technology Development (Project No. PJ009418022014 and PJ00911702014)" Rural Development Administration, Republic of Korea.

References

1. Willadsen SM (1986) Nuclear transplantation in sheep embryos. (Translated from eng). Nature 320(6057):63–65

2. Prather RS et al (1987) Nuclear transplantation in the bovine embryo: assessment of donor nuclei and recipient oocyte. (Translated from eng). Biol Reprod 37(4):859–866

3. Campbell KH, Loi P, Otaegui PJ, Wilmut I (1996) Cell cycle co-ordination in embryo cloning by nuclear transfer. (Translated from eng). Rev Reprod 1(1):40–46

4. Campbell KH, McWhir J, Ritchie WA, Wilmut I (1996) Sheep cloned by nuclear transfer from a cultured cell line. (Translated from eng). Nature 380(6569):64–66

5. Wilmut I, Schnieke AE, McWhir J, Kind AJ, Campbell KH (1997) Viable offspring derived from fetal and adult mammalian cells. (Translated from eng). Nature 385(6619): 810–813

6. Wells DN, Misica PM, McMillan WH, Tervit HR (1998) Production of cloned bovine fetuses following nuclear transfer using cells from a fetal fibroblast cell line. (Translated from English). Theriogenology 49(1):330

Chapter 4

Nuclear Transfer and Transgenesis in the Pig

Mayuko Kurome, Barbara Kessler, Annegret Wuensch, Hiroshi Nagashima, and Eckhard Wolf

Abstract

Somatic cell nuclear transfer (SCNT) using genetically modified donor cells facilitates the generation of tailored pig models for biomedical research and for xenotransplantation. Up to now, SCNT is the main way to generate gene-targeted pigs, since germ line-competent pluripotent stem cells are not available for this species. In this chapter, we introduce our routine workflow for the production of genetically engineered pigs, especially focused on the genetic modification of somatic donor cells, SCNT using in vitro matured oocytes, and laparoscopic embryo transfer.

Key words Pig, Cloning, Somatic cell nuclear transfer, Gene transfer, Gene targeting, Transfection, Embryo transfer

1 Introduction

Somatic cell nuclear transfer (SCNT) in combination with genetically modified donor cells is widely used for the generation of engineered large animals. Advantages of this strategy are—compared to conventional methods like pronuclear DNA microinjection or lentiviral gene transfer—that all produced offspring are genetically modified, which can be ensured by prescreening of the donor cells, and the obtained cloned animals show no mosaicism. In addition, cloned animals are genomic copies which may be an advantage for their use as large experimental animal models, because the genetic background is mostly not defined in livestock species.

During the last decade, transgenic pigs have gained importance in the field of biomedical research. Although basic research on disease mechanisms is mostly performed in rodent models, their use for translational studies is limited by size, lifespan, and physiology. The pig is an interesting alternative, showing close anatomical and physiological similarities to humans. Therefore, genetically tailored pig models play an increasingly important role for investigating disease

Nathalie Beaujean et al. (eds.), *Nuclear Reprogramming: Methods and Protocols*, Methods in Molecular Biology, vol. 1222, DOI 10.1007/978-1-4939-1594-1_4, © Springer Science+Business Media New York 2015

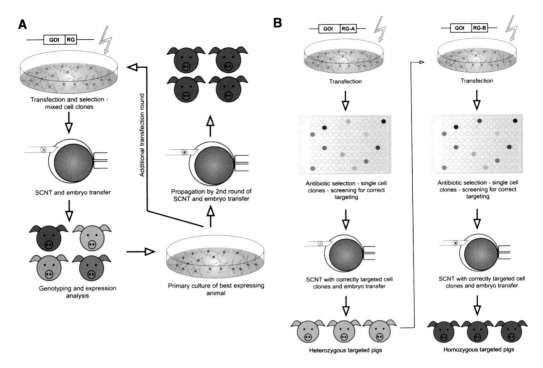

Fig. 1 Two strategies for the generation of transgenic pigs by somatic cell nuclear transfer (SCNT). (**a**) Additive gene transfer: After transfection of primary cells with a vector containing the gene of interest (GOI) and a resistance gene (RG) and antibiotic selection for 7–10 days, pools of mixed cell clones are used as donor cells for SCNT. Cloned fetuses or offspring are genotyped and gene expression analysis is performed. Different integration sites and copy numbers of the transgene in the transfected cells could lead to variations in gene expression in the cloned pigs/fetuses (*pink* = no expression, *light* to *dark blue* = low to high expression levels). Founder animals with appropriate expression levels are either propagated by serial nuclear transfer or breeding. Transgenic cells can be used for further transfection to generate pigs carrying multiple genetic modifications. (**b**) Gene targeting: After transfection, single cell clones are produced by performing the antibiotic selection on 96-well plates. After analysis of these cell clones correctly targeted ones can be used for SCNT. Homozygous pigs can be produced either by breeding of the heterozygous knockout pigs or an additional round of transfection of isolated primary cells from the heterozygous pigs or fetuses with a targeting vector containing another resistance gene

mechanisms, testing efficacy and safety of new therapies, and for the identification of biomarkers for companion diagnostics.

Two strategies for the generation of genetically engineered pigs by SCNT are shown in Fig. 1 [1]. For additive gene transfer, an expression vector for the gene of interest—linked to a selectable marker gene—is transfected into nuclear donor cells and integrates randomly into the genome (Fig. 1a). Pools of stable transfected cell clones are used for SCNT followed by transfer of the cloned embryos into synchronized recipients. This results in the generation of multiple founder animals, which have the same genetic background, but differ in transgene copy numbers and integration sites, and—consequently—in transgene expression levels and patterns. This provides the opportunity to select the cloned founder pig with the most appropriate transgene expression level/pattern for a particular experiment. Selected founders can be propagated by re-cloning or by breeding. In addition, cells from genetically modified founder fetuses/animals can be used for further transfection to generate pigs carrying multiple genetic modifications.

On the other hand, gene targeting is referred to as a site-specific modification by homologous recombination of the locus of interest, resulting in inactivation of the target gene ("knockout") or in a targeted insertion of a gene construct ("knock-in"; Fig. 1b). In contrast to additive gene transfer, it requires long-term in vitro culture of single cell clones and expansion to sufficient cell numbers for screening prior to SCNT. This long-term culture of somatic cells may induce changes which lead to a decrease in cloning efficiency. However, since germ line-competent pluripotent stem cells—the key to sophisticated reverse genetics in rodents—are not available in pig, SCNT is so far the main way to generate pigs with targeted genetic modifications. Homozygous knockout pigs are either generated by sequential targeting of both alleles of the target gene using different selectable marker genes or by breeding. Until now several genetically tailored pig models were successfully established by these strategies in our laboratories [2–6].

Although SCNT is known as a powerful tool for the generation of tailored pig models, various questions and problems are remaining. The cloning efficiency (percentage of cloned offspring out of transferred embryos) is still relatively low, usually ranging from 1 to 5 %. Additionally, several developmental defects can be observed in cloned fetuses/offspring. In our data set from the last few years, we observed that many cloned pigs were stillborn or died soon after birth [7]. The major reason for early neonatal death within 2 weeks was underweight and/or weakness of unknown causes, which was observed in several litters of piglets cloned from genetically modified cells. These phenotypic abnormalities could not be associated with any particular parameters, like donor cell source or genetic modification, and might be a side effect in pig cloning due to errors of epigenetic reprogramming.

In this chapter, we will describe detailed protocols of our routine workflow for the production of genetically engineered pigs. The first part covers the establishment of genetically modified donor cells, the second and third parts describe the technique of nuclear transfer using in vitro matured oocytes and the laparoscopic embryo transfer procedure.

2   Materials

Either purchased water or ultrapure water with a resistivity of 18.2 MΩ-cm is used for the preparation of all media (*see* **Note 1**). Sterilization of the media is done by using a 0.22 μm membrane filter (*see* **Note 2**).

2.1 Culture and Transfection of Donor Cells

1. Donor cells: porcine fetal fibroblasts (PFF), porcine ear fibroblasts (PEF), or porcine kidney cells (PKC) (*see* **Note 3**).

2. 0.02 % Collagen type 1 diluted with sterilized water. Prepare freshly before coating.

3. Dulbecco modified Eagle medium (DMEM) supplemented with 293 mg/L L-glutamine, 100 U/mL penicillin, 100 µg/mL streptomycin, 0.1 mM 2-mercaptoethanol, 1 % (v/v) non-essential amino acids, 1 % (v/v) sodium pyruvate, and 10–20 % fetal calf serum (FCS). After filtering, store at 4 °C (*see* **Note 4**).

4. Phosphate-buffered saline (PBS) without Ca^{2+} and Mg^{2+}.

5. Trypsin/EDTA: 0.4 % trypsin/0.032 % EDTA dissolved in PBS. After filtering, store at −20 °C until use.

6. 1–4 µg endotoxin-free prepared and linearized DNA (conventional plasmids or BAC vectors).

7. Nucleofection system and kit (*see* **Note 5**).

8. Antibiotic solution for selection: dissolve powder according to the manufacturer's protocol and store stocks at −20 °C or use ready-to-use solutions. Dilute with culture medium to the appropriate final concentration.

9. Cryopreservation medium: freshly prepared; 10 % dimethyl sulfoxide (DMSO) + 90 % FCS; store on ice until usage.

10. Standard tissue culture-treated dishes/plates.

11. 96-well half area or 96-well tissue culture-treated microplates.

12. Cell culture microscope and, if favored, an appropriate visualization and detection system for the screening of cell colonies (*see* **Note 6**).

13. Incubator: CO_2 controlled (5 % CO_2 in air), humidified atmosphere, at 37 °C.

2.2 Somatic Cell Nuclear Transfer

2.2.1 In Vitro Maturation of Pig Oocytes

1. Antibiotic stock (×100): 5 g/L streptomycin sulfate, 8 g/L penicillin-G, 25 mg/L amphotericin B and 153.8 mM NaCl in water. After sterilization, store at −20 °C.

2. PBS supplemented with 0.1 % (w/v) polyvinyl alcohol (PVA). Sterilize and store at room temperature for few weeks. PVA can be replaced by FCS. Add the antibiotic stock (×100) before use. The solution is used for transportation and washing of ovaries.

3. Cetyltrimethyl ammonium bromide (CETAB) stock: Mix 25 g CETAB in 125 mL ethanol (95 %), and make up to 500 mL with water. Store at room temperature (*see* **Note 7**).

4. CETAB working solution: Dilute CETAB stock to 0.2 % solution with PBS. Store at room temperature.

5. HEPES-buffered Tyrode's lactose medium with 0.1 % (w/v) PVA (TL-HEPES-PVA): 113.8 mM NaCl, 3.2 mM KCl,

2 mM NaHCO$_3$, 0.35 mM NaH$_2$PO$_4$, 10 mM sodium lactate, 2.0 mM CaCl$_2$, 0.5 mM MgCl$_2$, 10 mM HEPES, 0.027 mM phenol red, 5 mM glucose, 43.9 mM sorbitol, 50 mg/L streptomycin sulfate, 65 mg/L penicillin-G in water. Adjust pH to 7.2–7.4 with 5 N NaOH. Sterilize the medium using a filter unit (0.22 μm) and store it for up to 2–3 weeks at 4 °C. This medium is basically used during manipulating of oocytes/embryos outside the incubator. Keep the temperature at 38.5 °C in water bath during an experiment. TL-HEPES-PVA can be replaced by HEPES-buffered tissue culture medium 199 plus 10 % FCS.

6. North Carolina State University (NSCU)-23 medium [8]: 108.73 mM NaCl, 4.78 mM KCl, 1.70 mM CaCl$_2$, 1.19 mM KH$_2$PO$_4$, 1.19 mM MgSO$_4$, 25.07 mM NaHCO$_3$, 5.55 mM glucose, 7.0 mM taurine, 50 mg/L streptomycin sulfate, 65 mg/L penicillin-G in water. Add L-glutamine ×100 stock (final concentration: 1 mM) and Hypotaurine ×100 stock (final concentration: 5 mM) before use. Sterilize the medium using a filter unit (0.22 μm) and store for up to 2 weeks at 4 °C.

7. L-glutamine stock (×100): Prepare a 99.2 mM solution in NCSU23. Filter, aliquot, and freeze at –20 °C for up to 3 months.

8. Hypotaurine stock (×100): Prepare a 499.9 mM solution in NCSU23. Filter, aliquot, and freeze at –20 °C.

9. Cysteine stock (×1,000): Prepare a 600.0 mM solution in water. Filter, aliquot, and freeze at –20 °C.

10. Porcine Follicular Fluid stock (×10): Collect follicular fluid from ovaries (3–6 mm of follicles in diameter) by aspiration and centrifuge at 1,200×g for 60 min at 4 °C. Filter, aliquot, and freeze at –20 °C.

11. Epidermal Growth Factor stock (×1,000): Prepare a 10 μg/mL solution in NCSU23 medium containing 0.1 % bovine serum albumin (BSA). Filter, aliquot, and freeze at –20 °C (*see* **Note 8**).

12. Hormone stock (×100): Prepare 1,000 IU/mL equine chorionic gonadotropin (ECG) and 1,000 IU/mL human chorionic gonadotropin (HCG) in saline. Filter, aliquot, and freeze at –20 °C.

13. Oocyte maturation medium: NCSU23 medium supplemented with 0.6 mM cysteine, 10 ng/mL epidermal growth factor, 10 % (v/v) porcine follicular fluid, 10 IU/mL ECG and 10 IU/mL HCG. Prepare the maturation medium on the day of use, and place into incubator (set at 38.5 °C) at least 1 h before use for equilibration.

14. Thermo-container: to keep temperature at 38.5 °C during transportation of ovaries.

15. Plastic beaker: for washing ovaries.

16. Water bath.

17. Hot plate and block heater: to control temperature at 38.5 °C during aspiration.

18. 10 mL disposable syringe and 20-G needle: for aspiration of follicular fluid.

19. Mouth pipette and pulled capillaries.

20. Disposable sterilized petri dishes 100 mm, 35 mm and 15 mL tube: for collection and selection of oocytes.

21. Stereo microscope.

22. Incubators: CO_2 controlled (5 % CO_2 in air) and both CO_2 and O_2 controlled (5 % CO_2, 5 % O_2, and 90 % N_2), set up at 38.5 °C.

23. Mineral oil (*see* **Note 9**).

2.2.2 Nuclear Transfer

1. TL-HEPES-PVA: for washing and manipulation of oocytes/embryos during nuclear transfer. Keep the medium in a water bath at 38.5 °C during the experiment.

2. NCSU23 with bovine serum albumin (BSA): Add 4 mg/mL BSA and refilter. Equilibrate the medium in an incubator (set at 38.5 °C).

3. Hyaluronidase stock (×100): Prepare a 0.01 g/mL solution in TL-HEPES-PVA. Filter, aliquot, and freeze at –20 °C.

4. TL-HEPES-PVA supplemented with 10 % FSC: for checking the first polar body.

5. Demecolcine treatment medium: NCSU23 medium supplemented with 0.4 μg/mL demecolcine, 0.05 M sucrose and 4 mg/mL BSA. Equilibrate the medium in incubator (set at 38.5 °C).

6. Cytochalasin B (CB) stock (x 1,000): Prepare a 5 mg/mL solution in DMSO. Prepare 10 μL aliquots and freeze at –20 °C.

7. Enucleation medium: TL-HEPES-PVA supplemented with 0.4 μg/mL demecolcine, 5 μg/mL CB and 10 % (v/v) FCS. Enucleation is performed in a 4 μL droplet covered with mineral oil.

8. 10 % Polyvinylpyrrolidone (PVP) in PBS: to clean the micropipette for micromanipulation.

9. Fusion medium: 280 mM mannitol, 0.5 mM HEPES, 0.15 mM $MgSO_4$ and 0.01 % (w/v) PVA in water. Adjust pH to 7.2–7.4 with 5 N NaOH. Filter (0.22 μm) and store at 4 °C for up to 2 weeks. The osmolarity should be 290 mΩ.

10. Activation medium: 280 mM mannitol, 0.05 mM $CaCl_2$, 0.1 mM $MgSO_4$ and 0.01 % (w/v) PVA in water. Filter (0.22 μm) and adjust pH to 7.2–7.4 with sterilized 1 N NaOH. Store at 4 °C for up to 2 weeks. The osmolarity should be 300 mΩ.

11. Cytochalasine B (CB) treatment medium: Add CB stock (final concentration: 5 μg/mL) to NSCU23 with BSA. Equilibrate the medium in incubator (set at 38.5 °C).

12. Porcine zygote medium (PZM-5; *see* ref. 9): medium for in vitro culture.

13. Stereo microscope with heating stage.

14. Mouth pipette and pulled capillaries.

15. Incubators: CO_2 incubator and CO_2/O_2 incubator (set at 38.5 °C).

16. Water bath or block heater (set at 38.5 °C).

17. Inverted microscope with microscope heating stage.

18. Micromanipulators.

19. Micropipettes: 30 μm diameter needle with spike for enucleation, 20 μm diameter without spike for cell insertion, diameter of holding pipette is 150 μm.

20. Electric cell fusion generator with two electrode needles (for fusion) or a fusion chamber slide (electrodes 1.0 mm apart, for activation).

21. Disposable sterilized petri dishes (35 mm).

22. Mineral oil.

2.3 Embryo Transfer

1. Altrenogest.

2. Equine chorionic gonadotropin (ECG).

3. Human chorionic gonadotropin (HCG).

4. Intravenous catheters.

5. 10 % Ketamine hydrochloride.

6. 2 % Xylazine.

7. Skin disinfectant.

8. Operation table, swiveling to 45°.

9. Endoscopic equipment: Hopkins forward oblique telescope (30°, 6.5 mm), atraumatic forceps (7.0 mm), two trocars (6 and 7 mm), cold light fountain (150 W), fiberoptic light cable, metal catheter with mandrin (155 mm × 2.5 mm).

10. Stereo microscope with heating stage.

11. Flexible central vein catheter.

12. Surgical sutures or clamps for closing the wounds.

3   Methods

The methods include three crucial steps for the generation of genetically modified pigs: (1) generation of genetically modified primary cells using nucleofection, (2) somatic cell nuclear transfer using in vitro matured oocytes, and (3) laparoscopic embryo transfer.

3.1   Generation of Genetically Modified Donor Cells Using Nucleofection

3.1.1   General Cell Culture Procedures

1. Prewarm culture medium, PBS, and trypsin/EDTA at 37 °C before use.

2. Coat culture plates with collagen (40–120 μL per cm^2) for 3 h at room temperature. After removal of the remaining liquid and drying they are ready for use or can be stored at 4 °C.

3. Culture the cells at 37 °C in a humid atmosphere of 5 % CO_2 in air.

4. Harvest the cells by washing two times with PBS followed by incubation with 30–90 μL of 0.1–0.4 % trypsin/0.08–0.32 % EDTA (concentration depends on the cell type; *see* **Note 10**) per cm^2 for 1–6 min at 37 °C. Stop the reaction with culture medium. Wash the cells by centrifugation at $200 \times g$ for 5–10 min.

5. The appropriate antibiotic concentration for selection (additive gene transfer and gene targeting) is cell type dependent and has to be determined in advance: seed the cells with a range of different concentrations and evaluate cell death. The suitable concentration should result in 100 % dead cells after 7–9 days (*see* Fig. 2). Medium is changed every other day.

3.1.2   Transfection of the Cells Using Nucleofection

Nucleofection is done according to manufacturer's instructions. The nucleofection solution should be at room temperature for nucleofection.

1. Thaw and seed fibroblasts/kidney cells with a low passage number (P2–3) in culture medium on collagen-coated plates at a density of 1.5–2×10^4 cells per cm^2.

2. Detach the cells (70–90 % confluence) with trypsin/EDTA 1 day before nucleofection and seed 1.5–2×10^4 cells per cm^2 on a new coated plate (*see* **Note 11**).

3. For nucleofection the cells should be 70–80 % confluent and contain many cells in mitosis (*see* Fig. 3 and **Note 12**).

4. Before nucleofection, equilibrate culture medium at least 30 min in the incubator in a 6-well plate to seed the cells after nucleofection.

5. Harvest the cells with trypsin/EDTA (appropriate concentration depending on the cell type) and count them.

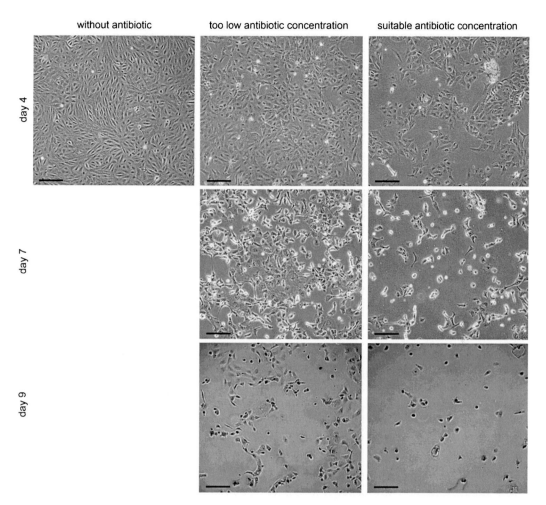

Fig. 2 Determination of appropriate antibiotic concentration for selection using nontransfected cells. The donor cells (here female porcine kidney cells) are treated with different antibiotic concentrations and evaluated during a period of 9 days. After 4 days an effect of the antibiotic treatment can be seen, dependent on the antibiotic concentration. The cells are growing slowly, are detaching, and their morphology is changing. After 7 days all cells (after treatment with a suitable antibiotic concentration) should be dead or in the process of dying. If there are still cell colonies on the dish the concentration of the antibiotic is too low. After 9 days all cells should be dead resulting in a nearly empty culture dish. Bars = 100 μm

6. Take $0.5–1.0 \times 10^6$ cells out of the suspension and centrifuge the cells (5 min, $200 \times g$).

7. Remove the supernatant, resuspend the cells in 100 μL nucleofection solution and mix it with 1–5 μg DNA.

8. Keep the cells in the nucleofection solution for a maximum of 15 min.

9. Transfer the suspension to the nucleofection cuvette included in the kit and put it into the nucleofector device. Choose the appropriate nucleofection program (*see* Fig. 4 and **Note 13**) and press the start bottom.

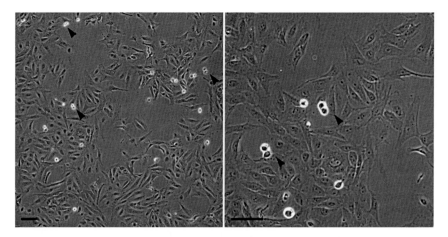

Fig. 3 Mitotic cells in a culture of female primary kidney cells. The success of transfection depends on the confluence of the target cells and amount of mitotic cells, especially in targeting experiments where single cell clones have to be generated and the demands on the cells are very high. Here, cells thawed at passage 2 and splitted 24 h before transfection are ~70 % confluent and exhibit an acceptable proportion of mitotic cells (marked by *arrowheads*). Bars = 50 μm

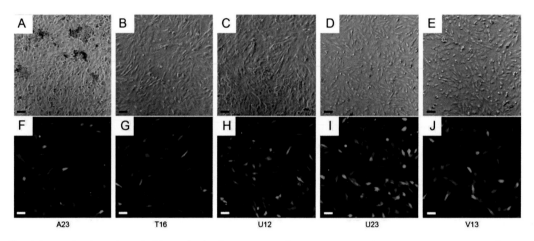

Fig. 4 Determination of a suitable nucleofection program. For each donor cell type appropriate nucleofection conditions have to be evaluated, using a GFP expressing plasmid, as shown here with male porcine fetal fibroblasts. Several programs (hereafter called A23, T16, U12, U23, V13) result in GFP expressing cells differing in percentage and intensity of fluorescent cells (increase from **f** to **j**), amount of detached cells (increase from **a** to **e**) and confluence (decrease from **a** to **e**). In this example, the program U12 is a good compromise regarding transfection efficiency and survival/quality of the nucleofected cells. Bars = 50 μm

10. Add 500–1,000 μL of culture medium (from the equilibrated 6-well plates) to the cuvette and slowly transfer the nucleofected cell suspension using the plastic pipette included in the nucleofection kit into the 6-well plate.

11. Keep the cells in the incubator for at least 24 h.

12. Change the culture medium 24 h after nucleofection.

13. Antibiotic selection can start 24–48 h after nucleofection.

3.1.3 Antibiotic Selection for Additive Gene Transfer

1. Remove the cells with trypsin/EDTA from the plate and count them.

2. Seed the cells into a 35–100 mm coated petri dish (depending on the cell number) in the appropriate selection medium. Use a higher seeding density if low transfection efficiency and few cell colonies are expected.

3. Culture the cells in selection medium for 7–10 days and change selection medium every other day.

4. When the cell colonies are confluent and 50–90 % of the dish is covered with cells (after 5–7 days depending on the transfection efficiency and population doubling time), harvest the cells and seed $1.5–2 \times 10^4$ cells per cm^2 on a new coated plate with selection medium. The cell colonies should not be confluent for a longer time period because of adhesion-dependent proliferative inhibition.

5. After a selection period of 7–10 days including one passaging step, stable transfected cells with 70–90 % confluence are ready for freezing.

6. Remove the cells with trypsin/EDTA, count, and centrifuge them (5 min, $200 \times g$).

7. Freeze the cells in cryopreservation medium at a controlled rate of -1 °C per min in a -80 °C freezer and transfer them for long-term storage to liquid nitrogen until somatic cell nuclear transfer.

3.1.4 Antibiotic Selection for Gene Targeting

1. Harvest the nucleofected cells with trypsin/EDTA and count them.

2. Seed 1×10^3 cells per well on a collagen-coated 96-well (half area) plate with selection medium (*see* **Notes 14** and **15**).

3. Change selection medium every other day.

4. After 7–9 days, all cells without integration of the transgene should be dead. The 96-well plates can then be screened for single cell colonies using a standard cell culture microscope or a system for automatic scanning and detection (*see* **Notes 16** and **17**).

5. Since single cell colonies are desired for further proceedings, it is important to detect wells containing only one colony (*see* Fig. 5).

6. Single cell colonies with an appropriate confluence are splitted to two 96-well standard tissue culture plates (*see* Fig. 5 and **Note 18**), one for DNA isolation (screening) and one for cryopreservation (donor cells for SCNT).

7. Harvesting the cells for DNA isolation: Let the cells grow until they are 90–100 % confluent. Detach the cells with suitable trypsin/EDTA concentration, stop the reaction using culture

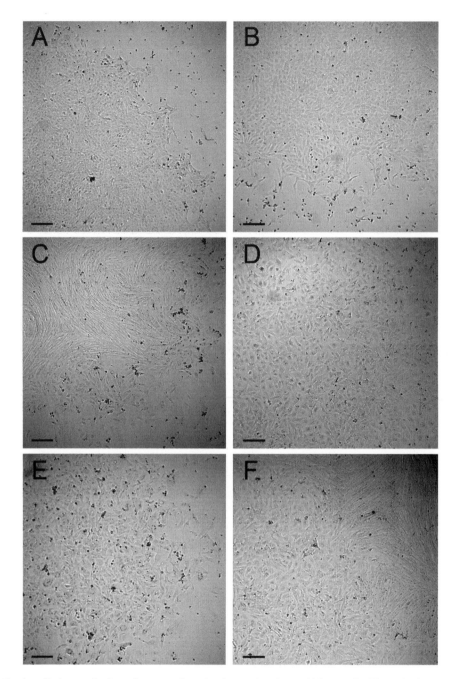

Fig. 5 Single cell clones of a targeting experiment using male primary kidney cells. The colonies may exhibit different cell morphologies, cell sizes, and population doubling times (**a–f**), since the primary culture is probably a mixed population of different cell types and transfection could have different effects on individual cells. If the cells of a colony are confluent for a longer time, adhesion-dependent proliferative inhibition could harm them and lead to a final proliferation stop, which could not be reversed (**c**). Some colonies can grow up to a certain size, but contain cells showing clear signs of senescence (enlarged, granula or vacuoles in the cytoplasm) and finally stop growing before reaching an appropriate cell number for splitting (**e**). Wells containing more than one cell colony (**f**) have to be discarded since single cell colonies are desired. Bars = 200 μm

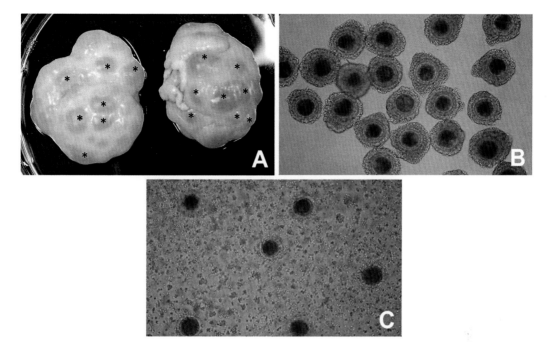

Fig. 6 Recovery and in vitro maturation of pig oocytes. (**a**) Ovaries obtained from a local abattoir. Only aspirate the follicles with a diameter of 3–6 mm (marked by asterisks). (**b**) COCs having at least 5–6 layers of compact cumulus cells are selected. Usually, per embryo transfer of about 100 embryos, 170–200 oocytes are cultured for maturation. (**c**) Oocytes with expanded cumulus cells after maturation for 2 days

medium and transfer the cell suspension to a reaction tube for centrifugation ($300 \times g$, 5–10 min). Freeze the pelleted cells after removal of the supernatant at –80 °C.

8. Cryopreservation: Detach the cells using suitable trypsin/ EDTA concentration, stop the reaction with 200 μL cryopreservation medium and transfer the suspension directly to a reaction tube which is stored at –80 °C for a short period until use.

3.2 Somatic Cell Nuclear Transfer

Manipulation of pig oocytes or embryos should be carried out on a heating stage at 38.5 °C during all steps until NT embryos are transferred to the recipient pig (*see* **Note 19**).

3.2.1 Preparation of In Vitro Matured Oocytes

1. Transport the ovaries obtained from a local abattoir in PBS with antibiotics at 38.5 °C using a thermo-container (*see* **Note 19**).

2. Wash the ovaries three times with CETAB working solution, and then rinse them three times with PBS with antibiotics. Keep the washed ovaries in PBS in a water bath (at 38.5 °C) until aspiration of follicles.

3. Aspirate antral follicles on the surface of ovaries using a 20-G needle attached to a 10-mL disposable syringe, and collect cumulus–oocyte complexes (COCs) with follicular fluid in a 15-mL tube, warmed in a block heater (set at 38.5 °C). Aspirate only follicles with a diameter of 3–6 mm (*see* Fig. 6a and **Note 20**).

4. Dilute the collected COCs plus follicular fluid with TL-HEPES-PVA, and select COCs which have at least 5–6 layers of compact cumulus cells under a stereo-microscope (*see* Fig. 6b and **Note 21**).

5. Wash selected COCs in three steps into maturation medium and culture them for 22 h in maturation medium with ECG and HCG. Then wash them into maturation medium without these hormones and culture for another 18–20 h. All cultures have to be done in a humidified atmosphere of 5 % CO_2, 5 % O_2, and 90 % N_2 at 38.5 °C (*see* **Note 22**).

3.2.2 Nuclear Transfer

1. Denudation of matured oocytes: After 2 days of maturation, briefly treat the oocytes containing expanded cumulus cells (Fig. 6c) with hyaluronidase, dissolved with TL-HEPES-PVA (0.1 mg/mL). After washing two times with TL-HEPES-PVA, remove cumulus cells from the oocytes by gentle pipetting with a finely drawn glass capillary pipette (internal diameter 150 μm, *see* **Note 23**).

2. Select oocytes displaying evenly granulated ooplasm and extrusion of the first polar body in TL-HEPES-PVA with 10 % FCS. Then culture for 0.5–1 h in demecolcine treatment medium in the incubator, to achieve a membrane protrusion near the first polar body, in which the condensed chromosome mass is located (*see* Fig. 7a and **Note 24**).

3. Enucleation of recipient oocytes on the inverted microscope equipped with heating stage and micromanipulators: Remove the chromosomes of recipient oocytes by aspirating the membrane protrusion, the first polar body and adjacent cytoplasm using a bevelled pipette with 30 μm diameter attached to the manipulator in TL-HEPES-PVA containing demecolcine (0.4 μg/mL), cytochalasin B (5 μg/mL) and 10 % FCS (*see* Fig. 7 and **Note 25**).

4. Preparation of donor cells: Thaw and culture donor cells until subconfluence, then starve the cells in culture medium containing 0.5 % FCS for 2–3 days (*see* **Note 26**). Shortly before use, cells are removed from the culture dish with trypsin/EDTA. Centrifuge the cell suspension ($200 \times g$, 5 min) and resuspend the pelleted cells in TL-HEPES-PVA.

5. Donor cell insertion into the enucleated oocyte: Select cells which are small and have a smooth cell membrane surface as donor cells (*see* Fig. 8A and **Note 27**). Insert a single donor cell into the perivitelline space of an enucleated oocyte using a bevelled pipette with 20 μm diameter in TL-HEPES-PVA with 10 % FCS (Fig. 8B). Store donor cell–oocyte complexes in the incubator until further use.

6. Fusion of donor cell–oocyte complexes: Wash donor cell–oocyte complexes two times with fusion medium, and place

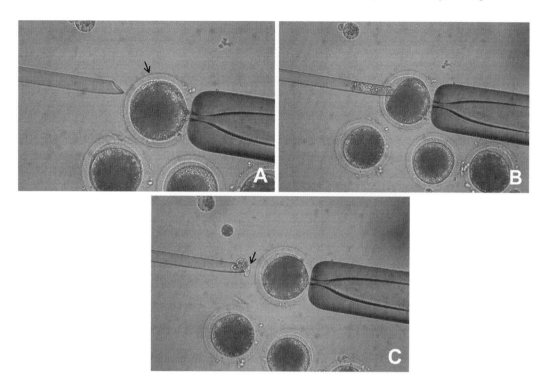

Fig. 7 Enucleation of recipient oocytes with a chemically assisted method. (**a**) After demecolcine treatment for 0.5–1 h, a membrane protrusion (*see arrow*), in which the condensed chromosome mass is located, can be observed near the first polar body. (**b**) The membrane protrusion, the first polar body and adjacent cytoplasm are removed by aspiration using a micromanipulator. (**c**) Removed membrane protrusion and the first polar body (*see arrow*)

about 20 donor cell–oocyte complexes in a 10-μL droplet of fusion medium (pH 7.2). Induce membrane fusion using an electric cell fusion generator and two electrode needles attached to micromanipulators by applying a single direct current (DC) pulse (200 V/mm, 20 μs × 1) and a pre- and post-pulse alternating current (AC) field of 5 V, 1 MHz for 5 s, respectively, to each single oocyte–donor cell complex (*see* Fig. 8c and **Note 28**).

7. Culture the reconstructed embryos in NCSU23 for 1–1.5 h before you start electrical activation.

8. Activation of NT embryos: Wash reconstructed embryos twice in activation solution, then place them between two wire electrodes (1 mm apart) of a fusion chamber slide overlaid with activation solution. Apply a single DC pulse of 150 V/mm for 100 μs (*see* **Note 29**).

9. Wash reconstructed embryos two times with TL-HEPES-PVA, and culture activated oocytes in CB treatment medium (5 μg/mL) for 3 h to suppress extrusion of the pseudo-second polar body.

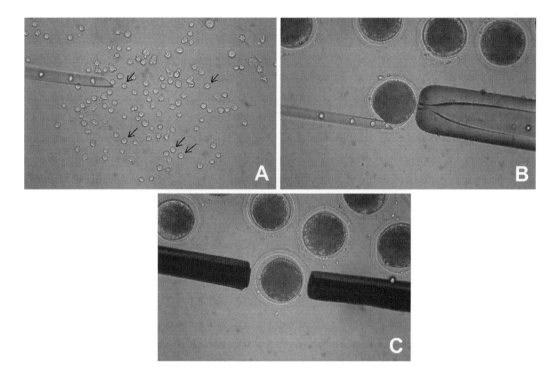

Fig. 8 Somatic cell nuclear transfer. (**a**) Selection of donor cells: Donor cells which show a small size and a smooth surface of the cell membrane (marked by *arrows*) are selected. (**b**) Donor cell insertion into the enucleated oocyte: A single donor cell is inserted into the perivitelline space of an enucleated oocyte. (**c**) Fusion of donor cell–oocyte complexes: Membrane fusion of the donor cell and the oocyte is induced by electric pulses using two electrode needles attached to micromanipulators

10. In vitro culture of NT embryos: Wash reconstructed embryos, and culture around 20 NT embryos in 20-μL droplets of porcine zygote medium 5 (PZM-5) under mineral oil in a plastic petri dish maintained in a humidified atmosphere of 5 % CO_2, 5 % O_2, and 90 % N_2 at 38.5 °C until embryo transfer.

3.3 Laparoscopic Embryo Transfer

1. Use gilts from 6 to 7 months of age as recipient pigs.

2. Administrate altrenogest orally for 15 days.

3.3.1 Synchronization of Recipients

3. Inject 750 IU ECG intramuscularly (i.m.) 24 h after last altrenogest administration.

4. Inject 750 IU HCG i.m. 80 h after ECG (*see* **Note 30**).

3.3.2 Embryo Transfer

1. Pigs should be fasted for 12 h before operation, and should be driven to the operation room at least 30 min before transfer, to stimulate urination.

2. Anesthesia of recipients: Administrate a combination of ketamine hydrochloride (1.2 mg/10 kg) and xylazine (0.5 mL/10 kg) by an intravenous ear catheter.

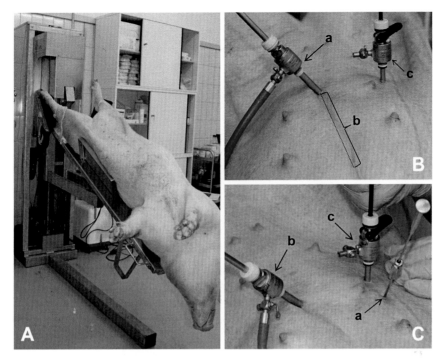

Fig. 9 Laparoscopic embryos transfer. (**a**) Fixation of recipient pig on an operation table in a 45° head-down position. (**b**) Position of a trocar for the optic (**a**) in the median about one hand after the umbilicus (*see* distance **b**), and a second trocar for atraumatic forceps (**c**). (**c**) Position of a hollow needle (**a**) to lead the vein catheter into the oviduct. Position of the optic and atraumatic forceps shown by **b** and **c**, respectively

3. Fix gilt in dorsal recumbence on an operation table and bring into 45° head-down position (*see* Fig. 9a).

4. Wash and disinfect abdomen.

5. Operation by laparoscopy: Place a trocar for the optic in the median about one hand after the umbilicus (*see* Fig. 9b).

6. Insert a 30° optic and control correct position in abdominal cavity.

7. Insufflate the abdomen with air or CO_2.

8. Place a second trocar for atraumatic forceps 10 cm caudolateral to the optic (*see* Fig. 9b).

9. Search ovary, identify transition from oviduct to infundibulum (*see* **Note 31**).

10. Collect infundibulum from ovary surface and fix it with forceps (*see* Fig. 10a).

11. Insert a hollow needle (length: 105 mm, inner diameter: 1.4 mm) through the abdomen into the oviduct (*see* Figs. 9c and 10a).

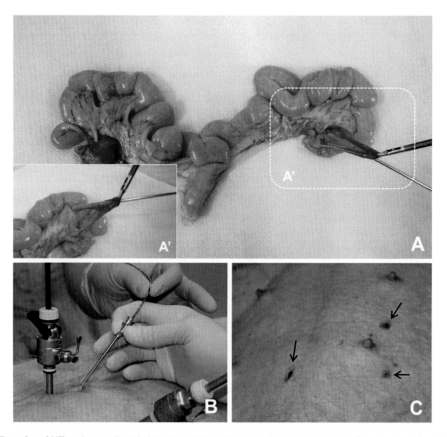

Fig. 10 Transfer of NT embryos directly into the oviduct by laparoscopy. (**a**) Uterus of a pig: the infundibulum is collected from the ovary surface and fixed with the forceps. The hollow needle is inserted directly into the beginning of the oviduct (*see* **a′**: another angle of the infundibulum). (**b**) The vein catheter loading NT embryos is inserted into the oviduct through the hollow needle. (**c**) Three small incisions remain after the endoscopic operation; far less tissue is damaged than after a surgical transfer

12. Collect NT embryos in TL-HEPES-PVA medium into a flexible central vein catheter (length 350 mm) under stereo-microscope with warming plate (*see* **Note 32**).

13. Insert vein catheter through the hollow needle about 4–5 cm into the oviduct and inject embryos (*see* Fig. 10c and **Note 33**).

14. Remove hollow needle and grasp, discharge remaining air in the abdomen through the trocars, remove trocars.

15. Close wounds by a clamp or stitch (*see* Fig. 10c and **Note 34**).

4   Notes

1. Quality of water critically affects the developmental competence of SCNT embryos.

2. When you prepare a small amount of medium (less than 10 mL), the first few droplets should be discarded to avoid filter cytotoxicity.

3. Suitable donor cells should have at least the following properties: (1) exhibit a correct and stable karyotype, which could be simply determined by counting the chromosome numbers in metaphase spreads, (2) show good proliferation, (3) have a long lifespan, and (4) should be transfectable. Furthermore, an in vitro SCNT experiment gives information about developmental competence until blastocyst stage.

4. For culturing fibroblasts from ear and fetus we use medium containing 15 % FCS for standard culture and 20 % after nucleofection. Kidney cells are normally cultured in medium containing 10 % FCS, which is increased to 15 % after nucleofection.

5. We recommend using the Nucleofector™ II device and the Amaxa™ Basic Nucleofector™ Kit Primary Fibroblasts (Lonza Cologne GmbH), or similar.

6. We recommend using the Cellavista System (SynenTec GmbH), or similar for screening and detection of single cell colonies.

7. Although the solution is clouded while mixing CETAB in ethanol, the solution becomes clear after addition of water.

8. Protein (BSA) is added to the solution, to prevent attachment of epidermal growth factor to the plastic surface.

9. Quality of mineral oil critically affects the cloning efficiency. Therefore, every new lot of oil should be tested before use.

10. We choose a trypsin/EDTA concentration resulting in detachment of the cells after 1–6 min. If the trypsin concentration is too high, leading to a very quick detachment, the cells could be harmed. Kidney cells are detached using 0.4 % trypsin/0.32 % EDTA, ear and fetal fibroblasts using 0.1 % trypsin/0.08 % EDTA.

11. We passage the cells after thawing to give the cells more time to recover after cryopreservation. It is also possible to transfect the cells after thawing and culture them until 70–80 % confluence without passaging.

12. Mitotic cells: Adherent fibroblasts/kidney cells change their morphology in mitosis, from spindle-shape to round, and nearly detach. These cells can be easily detected under the microscope (Fig. 3).

13. Determination of a suitable nucleofection program should be tested in advance, e.g. using a GFP-expressing plasmid. The following parameters can be evaluated: morphology, cell quality, amount of cells in suspension probably reflecting dead cells, cell confluence, percentage of transfected cells (indicated by fluorescent cells) and fluorescence intensity (Fig. 4). The best compromise between transfection efficiency and quality/survival of nucleofected cells has to be determined.

14. The 96-well half area plates are only necessary when scanning and screening of the plates is performed with the Cellavista system (SynenTec GmbH). The half area improves the optical characteristics by reducing the dark edges of the wells caused by the meniscus, allowing automatic detection of cell colonies by the Cellavista software. If the plates are screened using a simple cell culture microscope, then seed 2×10^3 cells per well on a standard tissue culture-treated 96-well plate.

15. High transfection efficiency could result in a relatively high proportion of wells containing more than one cell colony. These cell clones cannot be separated from each other and thus have to be discarded since single cell clones are desired for analysis. Repeat the experiment and reduce the number of nucleofected cells per well if many wells with more than one colony are detected after selection. Overall we seed ~1×10^3 cells per well containing nucleofected and nonnucleofected cells. The overall amount of cells seeded per well and the optimal ratio of nucleofected and nonnucleofected cells depend on the properties of the cell type used and have to be evaluated.

16. The perfect time point for screening may differ between different cell types and experiments. It therefore has to be determined by monitoring the selection progress.

17. Best results in colony detection are obtained with 100 µL medium per well on the 96-well half area plates. A higher volume improves the optical appearance by brightening the dark edges of the wells. This facilitates the detection of cell colonies if done with a cell culture microscope as well as with the Cellavista system. For the optimal detection of cell colonies by the Cellavista software some optimization of the preset templates is necessary.

18. The minimum density for cell splitting depends on the growth properties of the cells and on the area of the 96-well plate (half area or standard). Single cell clones may differ in their morphology, population doubling time, and life span (*see* Fig. 5a–f). On the half area plates, we split the cells when they cover 50–100 % of the area. Consider that cell colonies with 100 % confluence in their center stop proliferation and that it probably cannot be reversed by splitting (*see* Fig. 5c).

19. Pig oocytes and embryos are very sensitive to low temperature due to the high levels of cytoplasmic lipids. Therefore, keeping the temperature at 38.5 °C (at least over 30 °C) is the most critical factor for handling pig oocytes and embryos. The temperature during transportation of ovaries from the slaughterhouse to the laboratory should also be carefully controlled. Finally, oocytes and embryos should not be exposed to any unnecessary

extra light under the microscope during manipulation, and should be stored preferentially in an incubator between all the treatment steps.

20. Good quality oocytes, with at least 5–6 layers of compact cumulus cells predominantly can be collected from 3 to 6 mm follicles.

21. It is very critical to choose good COCs for the following experiment. COCs with parts of the cumulus cells missing or dead (color of the cumulus cells looks darker), as well as morphologically abnormal ones should be discarded.

22. In vitro maturation of COCs is performed in our laboratory in microdroplets: Around 20–25 COCs are cultured in a 100-μL droplet covered with mineral oil in a plastic 35-mm petri dish. Usually, per embryo transfer of about 100 embryos, 170–200 oocytes are cultured for maturation. Average percentage of maturation in our laboratory is around 80 %.

23. All cumulus cells should be removed; otherwise oocytes manipulation by micromanipulators will be difficult.

24. The chromosomes of a pig oocyte are not visible due to the high content of cytoplasmic lipids. In the classical "blind" enucleation, the first polar body and the underlying cytoplasm are aspirated. The enucleation efficiency can be improved by a chemically assisted method developed by Yin et al. [10], which leads to a membrane protrusion containing the chromosomes. Although dependent on the oocyte quality, the membrane protrusion can be observed in around 40–60 % of the matured oocytes in our laboratory.

25. Successful removal of the chromosomes can be verified by DNA staining with bisbenzimide (Hoechst 33342; 5 μg/mL) under a fluorescent microscope (350/461 nm). However, in the case of an experiment including an embryo transfer to get fetuses/animals, DNA staining is normally skipped because of the risk of DNA damaging.

26. Synchronization of the donor nucleus at G0/G1 is one of the key factors for successful nuclear transfer, because the chromosomal diploidy of the NT embryos must be maintained. Serum starvation is an effective synchronization method, showing around 85 % synchrony in G0/G1 for fibroblasts after 2–3 days of serum starvation [11]. Although a prolonged duration of serum starvation increased the population of G0/G1 cells, the potential risk of inducing apoptosis also increased. Therefore, the duration should be minimized to maintain cellular integrity of the donor cells.

27. It is assumed that relatively small-sized cells are G0/G1 stage, while bigger cells are G2/M stage.

28. In our laboratory, fusion is done by two electrode needles (NEPA GENE Co., Ltd.). The donor cell–oocyte complexes should be brought into one line between the two electrode needles attached to the micromanipulators [12]. Usually more than 90 % of the donor cell–oocyte complexes are fused by this technique.

29. The best activation time for the embryo is around 45 h after the beginning of oocyte maturation. When the reconstructed embryos are exposed to the activating electric pulse, no extra ions should be brought into the activation medium.

30. Embryo transfer can be performed either the same day of the experiment or 1–2 days after HCG injection.

31. Before embryo transfer, ovulation state of the recipients should be controlled. Ovaries should show either huge preovulation follicles or fresh ovulation cavities. The infundibulum must be treated very carefully to prevent bleeding and rupture.

32. NT embryos are put into a tube filled with NCSU23 plus BSA medium, and are transported to the operation room using a thermo-container (38.5 °C). NT embryos are picked up from the tube, and aspirated with TL-HEPES-PVA medium into a flexible central vein catheter between two air columns. Inject as little fluid and air as possible. Around 100 NT embryos are usually transferred to a recipient pig in our laboratory.

33. A unilateral transfer is sufficient due to the junction of both uterine horns in the pig.

34. A surgical transfer with laparotomic access to the abdomen and oviducts is possible as well, but the laparoscopic procedure is much faster (when done by a skilled person) and causes far less tissue damage in the recipient.

Acknowledgement

Our projects involving the generation of genetically engineered pigs are/were financially supported by the German Research Council (FOR 535 "Xenotransplantation," FOR 793 "Mechanisms of Fracture Healing in Osteoporosis," Transregio-CRC 127 "Biology of xenogeneic cell, tissue and organ transplantation—from bench to bedside"), by the Federal Ministry for Education and Research (Leading-Edge Cluster "m⁴—Personalised Medicine and Targeted Therapies"), the Bavarian Research Council (FORZebRA, Az. 802-08), and by the Mukoviszidose Institut gemeinnützige Gesellschaft für Forschung und Therapieentwicklung mbH. Authors are members of COST Action BM1308 "Sharing Advances on Large Animal Models – SALAAM".

References

1. Richter A, Kurome M, Kessler B et al (2012) Potential of primary kidney cells for somatic cell nuclear transfer mediated transgenesis in pig. BMC Biotechnol 12:84

2. Umeyama K, Watanabe M, Saito H et al (2009) Dominant-negative mutant hepatocyte nuclear factor 1alpha induces diabetes in transgenic-cloned pigs. Transgenic Res 18:697–706

3. Klymiuk N, Bocker W, Schonitzer V et al (2012) First inducible transgene expression in porcine large animal models. FASEB J 26: 1086–1099

4. Klymiuk N, Mundhenk L, Kraehe K et al (2012) Sequential targeting of CFTR by BAC vectors generates a novel pig model of cystic fibrosis. J Mol Med 90:597–608

5. Klymiuk N, van Buerck L, Bahr A et al (2012) Xenografted islet cell clusters from INSLEA29Y transgenic pigs rescue diabetes and prevent immune rejection in humanized mice. Diabetes 61:1527–1532

6. Renner S, Braun-Reichhart C, Blutke A et al (2013) Permanent neonatal diabetes in INS[C94Y] transgenic pigs. Diabetes 62:1505–1511

7. Kurome M, Geistlinger L, Kessler B et al (2013) Factors influencing the efficiency of generating genetically engineered pigs by nuclear transfer: multi-factorial analysis of a large data set. BMC Biotechnol 13:43

8. Petters RM, Wells KD (1993) Culture of pig embryos. J Reprod Fertil Suppl 48:61–73

9. Yoshioka K, Suzuki C, Tanaka A et al (2002) Birth of piglets derived from porcine zygotes cultured in a chemically defined medium. Biol Reprod 66:112–119

10. Yin XJ, Tani T, Yonemura I et al (2002) Production of cloned pigs from adult somatic cells by chemically assisted removal of maternal chromosomes. Biol Reprod 67:442–446

11. Tomii R, Kurome M, Ochiai T et al (2005) Production of cloned pigs by nuclear transfer of preadipocytes established from adult mature adipocytes. Cloning Stem Cells 7: 279–288

12. Kurome M, Fujimura T, Murakami H et al (2003) Comparison of electro-fusion and intracytoplasmic nuclear injection methods in pig cloning. Cloning Stem Cells 5:367–378

Chapter 5

Embryonic Stem Cell–Somatic Cell Fusion and Postfusion Enucleation

Huseyin Sumer and Paul J. Verma

Abstract

Embryonic stem (ES) cells are able to reprogram somatic cells following cell fusion. The resulting cell hybrids have been shown to have similar properties to pluripotent cells. It has also been shown that transcriptional changes can occur in a heterokaryon, without nuclear hybridization. However it is unclear whether these changes can be sustained following removal of the dominant ES nucleus. In this chapter, methods are described for the cell fusion of mouse tetraploid ES cells with somatic cells and enrichment of the resulting heterokaryons. We next describe the conditions for the differential removal of the ES cell nucleus, allowing for the recovery of somatic cells.

Key words Cell fusion, Reprogramming, Heterokaryons, Postfusion enucleation

1 Introduction

Somatic cells can be reprogrammed to a pluripotent state via cell fusion to embryonic stem cells. Cell fusion can be experimentally induced in a number of ways including using chemicals, such as PEG, electrical currents, and viruses [1], typically resulting in a tetraploid cell hybrid. In 1976 Miller and Ruddle were the first to show that the phenotype of the pluripotent cell type dominates, following cell fusion. They found that cell hybrids made between mouse Embryonal Carcinoma (EC) and thymocytes took on the pluripotent characteristics of EC cells and could form tumors containing derivatives of the three germ layers in a teratoma assay [2]. In most cell hybrids, the phenotype of the less-differentiated fusion partner dominates the phenotype of the more-differentiated partner; both embryonic germ cells and embryonic stem cells have been shown to reprogram somatic cells by cell fusion [3–6].

Reprogramming by cell fusion involves the hybridization of a somatic cell with a pluripotent cell resulting in a tetraploid cell hybrid. Even though ES-somatic cell hybrids display most properties of pluripotent stem cells, their tetraploid state renders them

Nathalie Beaujean et al. (eds.), *Nuclear Reprogramming: Methods and Protocols*, Methods in Molecular Biology, vol. 1222, DOI 10.1007/978-1-4939-1594-1_5, © Springer Science+Business Media New York 2015

incapable of significantly contributing to the late gestation epiblast [4] and chimeras [7]. Therefore, in order for cell fusion based reprogramming of somatic cells to be used as an alternative method of generating autologous pluripotent cells the removal of the ES cell derived DNA would be required following reprogramming [6, 8–10]. One approach that our laboratory has proposed involves the reprogramming of a somatic cell in a heterokaryon before nuclear fusion, followed by subsequent enucleation of the ES nucleus, which could potentially lead to the generation of autologous pluripotent cells [8, 9, 11].

In this chapter we describe the PEG-mediated cell fusion method using tetraploid ES cells and mouse mesenchymal stem cells (MSCs), maintain the fused cells as heterokaryons for 48–72 h and perform centrifugation based postfusion enucleation of the tetraploid nucleus.

2 Materials

2.1 Cells and Culture Media

1. Tetraploid mouse embryonic stem cells (4N hygro/puro-TK ES) a feeder free D3 strain, containing the thymidine kinase (TK) suicide gene generated by cell fusion between two clonal mouse ES D3 cell lines transfected with pSicoR-Hygro-Tk or pSicoR-Puro-Tk vectors [11], respectively. Cultured in ES cell medium containing 150 µg/mL Hygromycin and 1.5 µg/mL Puromycin.

2. Mesenchymal stem cells (MSCs) obtained from OG2 mice containing a GFP transgene under the control of the Oct4 promoter [11].

3. ES cell culture medium: DMEM supplemented with 10 % fetal bovine serum, 1 mM L-glutamine, 0.1 mM β-mercaptoethanol, 1,000 U/mL LIF, 1 % nonessential amino acids and 0.5 % Penicillin-Streptomycin.

4. Mesenchymal stem cell medium: α-MEM supplemented with 20 % fetal bovine serum.

5. Hygromycin B: 100 mg/mL stock in water.

6. Puromycin: 1 mg/mL stock in water.

2.2 Chemicals and Plastic Ware

1. Polyethylene glycol (PEG) MW 6000. Make a 50 % w/v solution in 150 mM HEPES, pH 7.5. Filter with 0.2 µm filter and store at −20 °C.

2. Cytochalasin B: 5 mg/mL stock in dimethyl sulfoxide (DMSO), stored at −20 °C.

3. Hoechst 33342: 5 mg/mL stock in water, stored at −20 °C away from light.

4. SNARF-1: 1 mg/mL stock in water for staining of somatic cells.

5. Ganciclovir: 10 mg/mL stock in 0.1 M HCl.

6. 6 well Tissue Culture Plates.

7. Bovine plasma cellular fibronectin-coated Enucleation wells/ disks. 1 cm tissue culture wells cut out from 4-well plates. Add 600 µL of cellular fibronectin, diluted at 10 µg/mL in sterile nanopure water, and let dry overnight under laminar flow.

8. 50 mL Neckless polycarbonate Sorvall centrifuge tubes, or equivalent.

9. Flexible plastic strips 1 cm wide × 7 cm long.

10. Cultures were maintained in a humidified incubator at 37 °C, with 5 % CO_2/95 % air.

2.3 Specialized Equipment

1. Sorvall high-speed centrifuge equipped with an Hb-4 swing out rotor, or equivalent.

2. FACS equipped with FITC and UV2a lasers.

3. Inverted phase contrast/epifluorescene microscope equipped with FITC and UV2A filters.

3   Methods

The method below describes the fusion of tetraploid mouse ES cells and SNARF-1-stained mouse mesenchymal stem cells (MSCs). The fused cells are sorted to enrich for heterokaryons (outlined in Fig. 1). The enriched heterokaryons are maintained in culture for 48–72 h before postfusion enucleation of the tetraploid nucleus (outlined in Fig. 2).

Day 1

1. In the morning plate the tetraploid mouse ES cells in mouse ES cell medium at a monolayer-like density for spin fusion.

2. The cells are to be plated at the following density; 2.5×10^6 per well of a 6-well plate. Return to incubator for overnight culture.

3. Preparation of MSCs for cell fusion. Take four confluent (75 %) T75 flasks of OG2 MSCs (10–20 million cells). Keep one flask of unstained cells for a FACs negative control.

4. The remaining two or three flasks are to be stained with SNARF-1 approximately 12–16 h before cell fusion is to be performed. Stain the MSCs with 1.5 µg/mL SNARF-1 directly in the culture flasks for 10 min.

5. Check for uptake of the stain under the fluorescence microscope.

6. Wash the cells in the flask twice with PBS (without Mg/Ca).

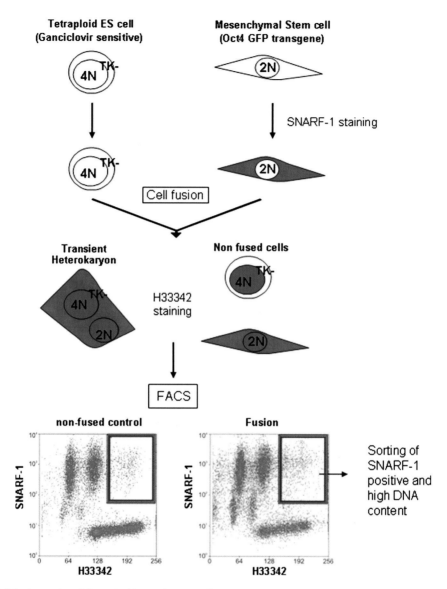

Fig. 1 Cell fusion and enrichment of heterokaryons. Schematic showing cell fusion between 4N hygro/puro-TK ES cells and OG2 MSCs. MSCs are prestained with SNARF-1 prior to cell fusion. Following cell fusion the cells are stained with H33342 and heterokaryons can be enriched by FACs by comparing profiles of nonfused control with the fused sample. Heterokaryons are enriched for by gating and sorting for SNARF-1 staining cells with high DNA content using H33342 staining

7. Add media and return to the incubator for overnight culture.

Day 2

8. Take out PEG 6000 and PBS (without Mg/Ca) and put in a 37 °C water bath.

9. Harvest MSCs, check for SNARF-1 (Red) staining of a few microliters with fluorescence microscopy compared with unstained controls.

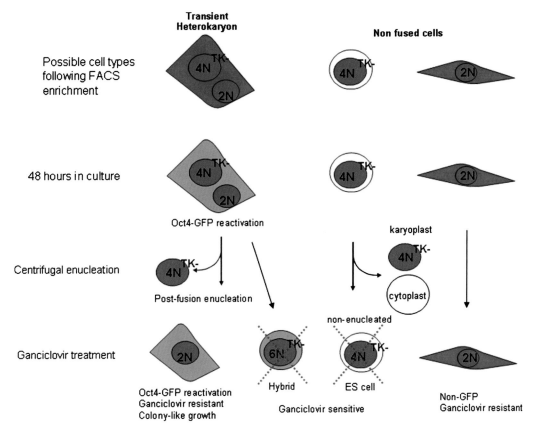

Fig. 2 Postfusion enucleation of heterokaryons. Schematic showing the possible cell types following FACs enrichment of heterokaryons. The outcomes following differential enucleation and ganciclovir treatment are shown. Tetraploid ES hygro/puro-TK ES cells as well as cell hybrids, which have not undergone enucleation are sensitive to ganciclovir, while MSC are insensitive and proliferate. Cells having undergone reprogramming can be enriched and identified by Oct4-GFP transgene expression and insensitivity to ganciclovir

10. Perform a cell count of the MSCs.

11. Keep a small amount of SNARF-1-stained cells for base line for FACs in media in tube at 37 °C (approx 200–500,000 cells), as well as unstained controls.

12. To perform spin fusions add 3×10^6 SNARF-1-stained MSCs onto the ES cells that were plated the day before in a 6-well plate.

13. Centrifuge the plate in a swing out rotor (SH-3000) at 2,000 rpm ($670 \times g$) at 20 °C for 10 min with an acceleration and deceleration setting of 7.

14. To perform cell fusions aspirate the media from the spun cells.

15. Add 1 mL of prewarmed PEG 6000 per 6 well gently by adding on to the side of the well.

16. Incubate for 2 min at room temperature and aspirate the PEG 6000 immediately.

17. Gently wash with 4 mL PBS (without Mg/Ca) and aspirate.

18. Repeat three times.

19. Add 3 mL ES media and allow cells to recover for as long as possible (2–3 h) before FACs. Also ensure to include a no-fusion no-PEG control for FACs.

20. After the cells have recovered from the spin fusions harvest the cells with trypsin, neutralize with media and centrifuge.

21. Aspirate the media and then stain the cells with H33342. Use 2 mL of media per 6-well fusion and stain with H33342 at a final concentration of 10 μg/mL.

22. Return the cells to the incubator for 45 min. Do not close the lid tight, flick intermittently.

23. In parallel harvest 4N ES cells and 2N MSCs cells. Stain these cells with H33342 for base line controls. Make sure to have a nonstained control as well.

24. After incubation wash off H33342 by adding 5 mL PBS (without Mg/Ca) and spin.

25. Repeat the wash.

26. Resuspend in 1 mL ES medium and transfer into FACS tubes and place on ice.

27. FACS analysis and sorting to be performed using a cell sorter, using the two parent cell lines and a mock-fusion control to set sorting gates. Remember to take a few empty FACS tubes and ES medium to sort into.

28. Set the base line auto fluorescence sorting gates by first analyzing unstained MSCs and 4N ES cells.

29. Next determine the sorting gates by analyzing red fluorescence using the SNARF-1-stained MSCs and then identify high DNA content ≥6N by FACS analysis of H33342-stained 4N ES cells.

30. Once the gates have been determined for SNARF-1 positive cells and high DNA content of ≥6N sort the fused sample for double positive cells into a fresh tube of mouse ES cells.

31. Plate at up to 100,000 sorted cells per cellular fibronectin-coated 1 cm enucleation wells. Also plate out control cells including MSCs and 4N ES cells. Return cells to the incubator for 48 h.

Day 3

32. Observe cell using phase-contrast or epifluorescence on an inverted microscope and return to incubator.

Day 4

33. To prepare the cells for postfusion enucleation place the wells upside down in 10–20 mL MEF media containing 10 μg/mL Cytochalasin B 50 mL in the neckless polycarbonate Sorvall

centrifuge tubes and incubate at 37 °C for 60 min. Make sure to treat the control wells also, including 4N ES cells and MSC alone (*see* **Note 1**).

34. To enucleate place the tubes in the HB-6 rotor in the Sorvall centrifuge and spin at 11,000 rpm (19,770×*g*) for 10 min at 20 °C.

35. Remove the wells from the tubes and replace the media with fresh ES medium added containing H33342 at 10 mg/mL and incubated for 30 min.

36. Observe and image the cells using an epifluorescence inverted OlympusIX71 microscope to determine enucleation rate and return to incubator for further culture. An example of heterokaryons before and after postfusion enucleation is shown in Fig. 3.

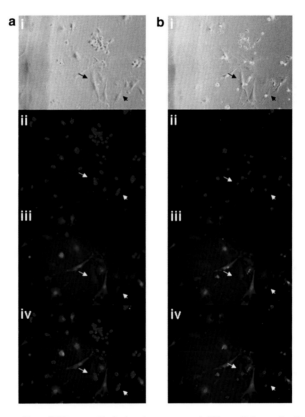

Fig. 3 Generation of ES-somatic heterokaryons and differential enucleation of 4N ES nucleus. (**a**) Heterokaryons (*arrows*) identified by (*i*) phase contrast, (*ii*) containing two nuclei visualized following H33342 staining and (*iii*) SNARF-1 staining, and (*iv*) merged image. (**b**) The same cells (*arrows*) identified after centrifugation and showing enucleation of the one nucleus (*i*) phase contrast, (*ii*) H33342 staining (*iii*) SNARF-1 staining, and (*iv*) merged image. Note that the other SNARF-1 negative cells which are tetraploid ES cells in the field of view which contained single nuclei were completely enucleated resulting in cytoplasts (enucleated cell cytoplasm), while SNARF-1 positive MSCs containing a single diploid nucleus do not undergo enucleation

Day 5

37. Aspirate the media and refresh with selection media containing 2 μM ganciclovir to select against the 4N hygro/puro-TK ES cells and culture for 5–7 days refreshing the media daily.

Days 10–12

38. Pick and expand any GFP positive colonies activating the Oct4-GFP transgene, or any other colonies of interest, for further analysis (*see* **Note 2**).

4 Notes

1. Enucleation is performed by placing 1 cm tissue culture wells into 50 mL Neckless polycarbonate Sorvall centrifuge tubes containing enucleation media, or equivalent. Flexible plastic strips 1 cm wide × 7 cm long are used to keep the wells in place and upside down at the bottom of the tube (*see* Fig. 4).

2. Further analysis of the resulting colonies is suggested. In this protocol we describe the use of ES and somatic cells from different genetic backgrounds, ESD3 129 strain and B6, respectively. Therefore, microsatellite DNA polymorphism analysis can be employed to confirm enucleation of the tetraploid nucleus and DNA using the following microsatellite marker primers.

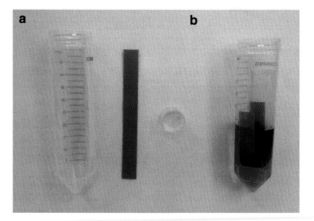

Fig. 4 Set up process for differential enucleation. (**a**) Requirements for centrifugal enucleation are neckless polycarbonate Sorvall centrifuge tubes, cell growing on cellular fibronectin-coated 1 cm culture wells, a sterilized flexible plastic strip. (**b**) Wells containing the cells are placed upside down in 10–20 mL MEF media containing 10 μg/mL Cytochalasin B 50 mL in the neckless polycarbonate Sorvall centrifuge tubes, or equivalent, and are held in place using a sterilized flexible plastic strip. This ensures the wells are kept upside down during centrifugation and remain completely horizontal

D1Mit206, D2Mit493, D3Mit259, D4Mit178, D5Mit246, D6Mit102, D7Mit31, D8Mit242, D9Mit311, D12Mit270, D13Mit3, D14Mit109, D16Mit57, D19Mit10, and DX170, obtained from the Broad Institute (http://www.broad.mit.edu).

Acknowledgements

We wish to acknowledge funding support from Dairy Australia, funding from the Dairy Futures CRC, and MIMR infrastructure funding supported by the Victorian Government's Operational Infrastructure Support Program.

References

1. Lucas JJ, Terada N (2003) Cell fusion and plasticity. Cytotechnology 41:103–109
2. Miller RA, Ruddle FH (1976) Pluripotent teratocarcinoma-thymus somatic cell hybrids. Cell 9:45–55
3. Tada M, Tada T, Lefebvre L, Barton SC, Surani MA (1997) Embryonic germ cells induce epigenetic reprogramming of somatic nucleus in hybrid cells. EMBO J 16: 6510–6520
4. Tada M, Takahama Y, Abe K, Nakatsuji N, Tada T (2001) Nuclear reprogramming of somatic cells by in vitro hybridization with ES cells. Curr Biol 11:1553–1558
5. Kimura H, Tada M, Nakatsuji N, Tada T (2004) Histone code modifications on pluripotential nuclei of reprogrammed somatic cells. Mol Cell Biol 24:5710–5720
6. Cowan CA, Atienza J, Melton DA, Eggan K (2005) Nuclear reprogramming of somatic cells after fusion with human embryonic stem cells. Science 309:1369–1373
7. Ying QL, Nichols J, Evans EP, Smith AG (2002) Changing potency by spontaneous fusion. Nature 416:545–548
8. Pralong D, Mrozik K, Occhiodoro F, Wijesundara N, Sumer H, Van Boxtel AL, Trounson A, Verma PJ (2005) A novel method for somatic cell nuclear transfer to mouse embryonic stem cells. Cloning Stem Cells 7: 265–271
9. Pralong D, Trounson A, Verma PJ (2006) Cell fusion for reprogramming pluripotency: towards elimination of the pluripotent genome. Stem Cell Rev 2:331–340
10. Yu J, Vodyanik MA, He P, Slukvin II, Thomson JA (2006) Human embryonic stem cells reprogram myeloid precursors following cell-cell fusion. Stem Cells 24:168–176
11. Sumer H, Jones KL, Liu J, Rollo BN, van Boxtel AL, Pralong D, Verma PJ (2009) Transcriptional changes in somatic cells recovered from embryonic stem-somatic heterokaryons. Stem Cells Dev 18:1361–1368

Chapter 6

Analysis of Nuclear Reprogramming Following Nuclear Transfer to *Xenopus* Oocyte

Jerome Jullien

Abstract

Germinal vesicle of stage V–VI *Xenopus Laevis* oocytes (at the prophase I stage of meiosis) can be used to transplant mammalian nuclei. In this type of interspecies nuclear transfer no cell division occurs and no new cell types are generated. However, the transplanted nuclei undergo extensive transcriptional reprogramming.

Here, it is first explained how to carry out transplantation of multiple mammalian cell nuclei to *Xenopus* oocytes. It is then described how to perform RT-qPCR, Western Blot, Chromatin Immunoprecipitation, and live imaging analysis to monitor transcriptional reprogramming of the nuclei transplanted to oocytes.

Key words *Xenopus*, Oocyte, Transcriptional reprogramming, Nuclear transfer, Germinal vesicle

1 Introduction

Cloned animals can be obtained from various species by transplanting somatic nuclei to an enucleated egg. During this process the somatic nucleus is reprogrammed so that it can give rise to all the different cell types needed to generate a whole organism. To improve the efficiency of nuclear transfer it is important to understand the mechanism of nuclear reprogramming. For that purpose alternative routes of nuclear transfer have been developed that facilitate the analysis of reprogramming [1]. One such a route is the nuclear transfer of mammalian nuclei to *Xenopus* oocyte (in prophase I of meiosis). In this type of interspecies nuclear transfer no cell division occurs and no new cell types are generated. However, the transplanted nuclei undergo extensive transcriptional reprogramming. This type of assay has been successfully used to identify maternal factors that are required for gene reactivation during nuclear transfer (histone variants, nuclear actin, gadd45) [2–5] as well as the component of somatic chromatin that resist the reprogramming process (macroH2A) [6]. In the following section I describe the

Nathalie Beaujean et al. (eds.), *Nuclear Reprogramming: Methods and Protocols*, Methods in Molecular Biology, vol. 1222, DOI 10.1007/978-1-4939-1594-1_6, © Springer Science+Business Media New York 2015

methods to transplant mammalian nuclei to *Xenopus* oocyte and to monitor transcriptional reprogramming by quantitative reverse transcription PCR (RT-qPCR), Western Blot analysis, Chromatin Immunoprecipitation, and live imaging.

2   Materials

2.1   Chemicals and Reagents

1. MS222: Ethyl 3-aminobenzoate methanesulfonate, 120 mg in 300 μl of water.

2. MBS (Barth-Hepes Saline) 10× stock : 88 mM NaCl, 1 mM KCl, 2,4 mM NaHCO$_3$, 0.82 mM MgSO$_4$, 0.33 mM Ca(NO$_3$)$_2$, 0.41 mM CaCl$_2$, 10 mM Hepes. Add ~3 ml of 10 N NaOH and check pH that should be at 7.4–7.6.

3. Liberase: reconstitute 100 mg of lyophilized enzymes with 10 ml sterile H$_2$O, giving a final concentration of 26 U/ml (we usually use the Liberase TM Research grade from Roche). Let the solution stand for 30 min on ice to rehydrate. Gently swirl the vial every few minutes to ensure that the enzyme is completely dissolved. Store aliquots of 250 μl (6.5 U) at –80 °C.

4. Penicillin Streptomycin (P/S): 10 mg/ml in water (1,000× stock).

5. Gentamicin: 10 mg/ml in water (1,000× stock).

6. Bovine Serum Albumin (BSA).

7. 0.25 % Trypsin-EDTA.

8. Phosphate Buffer Saline (PBS): 137 mM NaCl; 2.7 mM KCl; 10 mM Na$_2$HPO$_4$; 2 mM KH$_2$PO$_4$; pH adjusted to 7.4 with HCl.

9. Streptolysin O (SLO): 25,000 U of SLO are resuspended in 1.25 ml PBS 0.01 %BSA. 62.5 μl of dTT 100 mM is then added to the solution and incubated 2 h at 37 °C. Aliquots of 25 μl are stored at –80 °C.

10. SuNaSP : 250 mM Sucrose, 75 mM NaCl, 0.5 mM Spermidine, 0.15 mM Spermine.

11. 0.4 % trypan blue solution: 0.4 g trypan blue diluted in 100 ml of PBS 1×.

12. Mineral oil, embryo tested.

2.2   Material Required for Oocyte Microinjection and Nuclear Transfer

1. 3–10-years-old Female *Xenopus laevis*.

2. Disposable consumables: 10 cm diameter Petri dishes, 50 ml Falcon tubes, 1.5 ml Eppendorf tubes, Pasteur pipettes, 50 μm filters.

3. Small lab equipment: rocker for gentle agitation, vortex, hemacytometer, bench microcentrifuge, micropipettes and tips, marker pen, standard dissecting microscope, dissecting scissors, and forceps.

4. Drummond injector (Drummond Nanoject, Drummond Scientific Company) and injection needles (7″ Drummond #3-000-203-G/XL glass capillaries) (*see* **Note 1**).

5. Donor cells such as mouse embryonic fibroblasts (MEF) or embryonic stem cells (ES) with their respective cell culture medium.

6. mRNA encoding a protein of interest for microinjection (*see* **Note 2**).

2.3 Supplementary Material for Transcriptional Analysis

1. Sonicator equipped with an ultrasonic bath.

2. GV isolation buffer: 20 mM Tris–HCl, pH 7.5, 0.5 mM $MgSO_4$, 140 mM KCl.

3. Nondenaturing washing buffer: 150 mM Tris–HCl, pH 7.5, 50 mM NaCl, 150 mM KCl.

4. Homogenization buffer : 50 mM Hepes, pH 7.8, 140 mM NaCl, 1 mM EDTA, 1 % Triton X100, 0.1 % Na-deoxycholate.

5. Proteases inhibitors 1,000× stock solution: 104 mM AEBSF, 80 μM Aprotinin, 4 mM Bestatin, 1.4 mM E-64, 2 mM Leupeptin, 1.5 mM Pepstatin A.

6. 37 % Formaldehyde in water, histology grade.

7. Glycine for molecular biology >99.0 % (RT).

8. Dithiothreitol (DTT) for molecular biology >99.5 % (RT).

9. Sodium Dodecyl Sulfate (SDS) for molecular biology >98.5 % (GC).

2.4 Live Imaging Setup

1. Imaging slides with 8 mm diameter wells; 10 mm square coverslips #1.5.

2. Confocal microscope equipped for live imaging.

3 Methods

3.1 Oocyte Microinjection and Nuclear Transfer

3.1.1 Removal of Follicular Cells from Oocytes

1. Ovaries are removed from a female *X. laevis* terminally anesthetized by subcutaneous injection of MS222.

2. The ovaries are placed in a Petri dish containing 1× MBS.

3. Ovaries are then separated into small clumps (containing roughly 50–100 oocytes) using forceps (Fig. 1a).

4. 3–5 ml of such clumps are transferred to a 50 ml falcon tube that is then filled to 12.5 ml with 1× MBS.

5. 6.5 U of liberase are then added to the solution and the oocytes are incubated 2 h 20 min at room temperature (22 °C) with gentle agitation on a rocker (*see* **Note 3**).

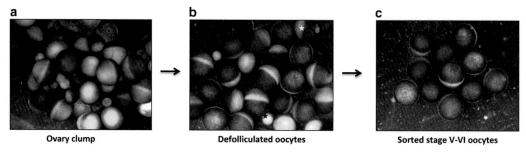

Fig. 1 Isolation of defolliculated oocytes from whole *Xenopus laevis* ovaries. (**a**) Ovaries isolated from a female *Xenopus laevis* are separated into small clumps using dissecting forceps. Oocytes at different stages of oogenesis are held together by a mesh of follicular cells and blood vessels. The ovary contains oocytes from all stages of oogenesis: from immature stage I oocytes (small and transparent) to mature stage V–VI oocyte (largest oocyte with a pigmented brown animal pole and a nonpigmented vegetal pole) (Fig. 3a). (**b**) After treatment with an enzymatic cocktail (liberase), the follicular cells are lost and individual oocytes are released. These defolliculated oocytes can then be sorted so that immature oocytes (*black stars*) or damaged stage VVI oocyte (*white star*) are removed. (**c**) After enzymatic defolliculation and sorting, stage V–VI oocytes are ready for the nuclear transplantation procedure

6. Defolliculated oocytes are then washed 4 times in 50 ml of 1× MBS and transferred to a Petri dish (Fig. 1b).

7. Individual oocytes are then manually sorted under a standard dissecting microscope using a Pasteur pipette and stored over night at 16–18 °C in 1× MBS supplemented with penicillin streptomycin (P/S) at a final concentration of 10 μg/ml.

3.1.2 mRNA Injection in Oocytes

For experimental purposes it is often required that the oocytes are injected prior to nuclear transfer with mRNA encoding a protein of interest. For efficient protein expression, the mRNA is injected 1 day before nuclear transplantation.

1. Using a Drummond injector, the required amount of mRNA is injected in the cytoplasm from the vegetal pole of the recipient oocyte (*see* **Note 4**).

2. mRNA injected oocytes are cultured in 1× MBS, P/S, 0.2 %BSA.

3. Oocytes can be kept for several days at 16–23 °C before undertaking nuclear transplantation.

3.1.3 Donor Cell Preparation

1. Adherent cells such as mouse embryonic fibroblasts (MEF) or embryonic stem cells (ES) are collected by trypsinization according to standard tissue culture procedure.

2. Trypsin treatment is stopped by addition of cell culture medium supplemented with serum.

3. The cells are then washed twice in 1× PBS by centrifugation for 3 min at $500 \times g$.

4. Prior to proceeding to permeabilization, single cells are collected by filtering the cell suspension through a 50 μm filters.

5. Cells are then counted on a hemacytometer and diluted to one million cells per ml.

6. One million cells (in 1 ml of 1× PBS) are transferred to an Eppendorf tube and centrifuged 2 min at 2,000 ×g at RT.

7. The cell pellet is then resuspended in 1 ml of SuNaSP.

8. The cells are centrifuged again and resuspended in 200 μl of SuNaSP.

9. 25 μl of SLO (500 U) is then added to the suspension and the cell are incubated 5 min at 37 °C in a water bath (*see* **Note 5**).

10. After incubation cells are placed on ice while the efficiency of permeabilization is checked. For that purpose 2 μl of the permeabilized cell suspension is mixed with 8 μl of 0.4 % trypan blue solution. Cells in trypan blue are visualized on a hematocytometer. Permeabilization is successful when more than 90 % of the cells show a nucleus stained blue and blebs are apparent on the cell plasma membrane (*see* **Note 6**).

11. When permeabilization is confirmed to be satisfactory by the trypan blue test, SLO is inactivated by adding 1 ml of SuNaSP 3 %BSA to the 225 μl of permeabilized cells.

12. Following SLO inactivation, cells are centrifuged 2 min at 2,000 ×g at RT.

13. The permeabilized cells pellet is then resuspended in 55 μl of SuNaSP-BSA to give a final concentration of 250 nuclei/13.8 nl of solution.

14. The permeabilized cells ("nuclei") are then used for transplantation without further manipulation (*see* **Note 7**).

3.1.4 Transplantation Procedure

Transcriptional reprogramming happens only when the nuclei are transplanted into the recipient oocyte own nucleus, the germinal vesicle (GV) [7].

1. First prepare the injection needle: its tip needs to be cut to give a relatively large opening (~50 μm) to be able to suck the donor nuclei suspension into the needle. This can be done simply by using a dissecting scissor positioned at a 120° angle to the needle. This will give a sharp enough needle tip that will penetrate the oocyte without damaging it.

2. It then is helpful to mark the needle with a marker pen at a distance of about 300 μm from the tip (Fig. 2b) (*see* **Note 8**).

3. The needle prepared in this way is filled with about 500 nl of the nuclei suspension (Fig. 3c).

4. To inject the nuclei, the oocyte is positioned towards the needle with forceps. The needle has to contact the middle of the pigmented (animal) pole of the oocyte and be orientated parallel to the animal/vegetal pole axis (Fig. 3e).

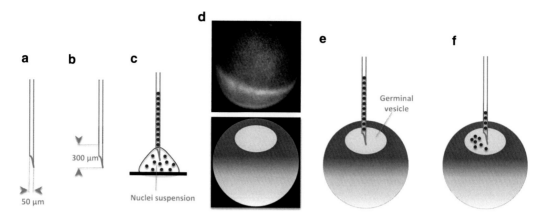

Fig. 2 Transplantation of nuclei to the germinal vesicle of *Xenopus laevis* oocytes. (**a**) The tip of a glass capillary is sharpened to an opening of 50 μm. (**b**) Using a marker pen, the needle is marked at about 300 μm from the tip. (**c**) Nuclei are gently sucked into the needle. (**d**) Picture of an oocyte with the animal pole (*brown pigmented layer*) facing up (*top*). Schematic representation of an oocyte cross-section showing the location of the oocyte germinal vesicle under the pigmented layer of the animal pole (*bottom*). (**e**) The needle is inserted in the middle of the animal pole of the oocyte, up to the marked area. (**f**) Nuclei are injected into the germinal vesicle

5. Once positioned, the needle is inserted in the oocyte at a depth of about 250 μm (using the pen mark as a reference) and the nuclei can then be injected (13.8 nl containing 250 nuclei) (*see* **Note 8**).

6. GV targeting success rate has to be checked for each injection needle used since slight differences in the size of the oocyte and the sharpness of the needle affect the injection depth required. This is done by isolating the GV from the oocyte in MBS 1× and verifying the presence of nuclei in the GV (Fig. 3). If the nuclei are not transplanted to the GV, the targeting test is repeated varying the injection depth, using the marker pen reference as a guide.

7. The nuclei suspension is kept on ice and the needle is refilled with nuclei from the stock solution every 20–30 oocytes injected (*see* **Note 9**).

8. After transplantation, the oocytes are cultured in 1× MBS supplemented with antibiotics (penicillin/streptomycin, gentamicin) and 0.2 % BSA.

9. The oocytes are kept at temperature of 16–23 °C and the culture medium is replaced every day.

10. Transcriptional reprogramming is typically assessed between 0 and 3 days after transplantation, but the transplanted oocytes can be cultured for a period of up to several weeks.

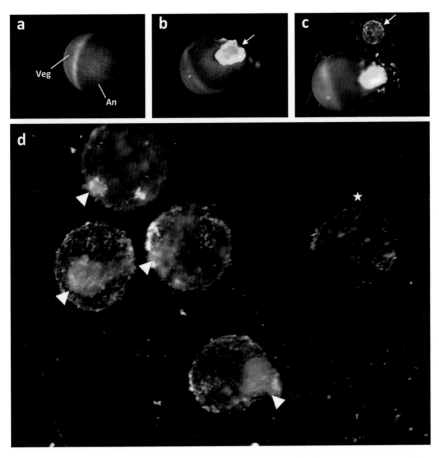

Fig. 3 Assessment of germinal vesicle targeting efficiency (**a**) Oocyte after nuclear transplantation to the germinal vesicle. The animal (An) and vegetal (Veg) poles of the oocyte are indicated. (**b**) The membrane at the animal pole is split open with dissecting forceps so that the underlying germinal vesicle is revealed (*arrow*). (**c**) The germinal vesicle is then pushed aside, free of the surrounding yolk granules (*arrow*). (**d**) Dissected germinal vesicles can be observed at high magnification under oblique light. The germinal vesicle successfully injected shows a white "cloud" corresponding to the transplanted nuclei (*arrowhead*), whereas the failed nuclear transfer results in a transparent germinal vesicle devoid of transplanted nuclei (*star*)

3.2 Preparation of the Samples for Transcriptional Analysis

When performing interspecies nuclear transfer (i.e. mouse nuclei to *X. laevis* oocyte) it is possible to monitor gene reactivation from transplanted nuclei by RTqPCR with qPCR primers that will specifically recognize transcripts from transplanted nuclei [8].

3.2.1 Analysis of Transplanted Nuclei by RT-qPCR Analysis

1. Pool of 4–8 oocytes are collected in Eppendorf tubes and can be stored at −80 °C or directly processed for mRNA extraction.

2. Alternatively, as it is possible to verify that all the collected oocytes contain transplanted nuclei properly targeted to the GV (*see* **step 6** in Subheading 3.1.4), the oocyte cytoplasm plus the GV containing the nuclei can then be directly aspirated

with a micropipette in a total volume of 50 μl, and transferred to an Eppendorf tube on dry ice (*see* **Note 10**).

3. mRNA extraction and qRT-PCR can then be performed using standard protocols

3.2.2 Analysis by Western Blot

After transplantation it is possible to recover nuclei from the oocyte GV. Using the technique described below it is possible to monitor by WB global changes occurring on transplanted chromatin during transcriptional reprogramming [9].

1. The GV containing transplanted nuclei are dissected out of the oocyte in GV dissection medium.

2. Within 1–2 min following dissection, the isolated GVs are transferred on ice to an Eppendorf tube containing nondenaturing wash buffer supplemented with proteases inhibitors.

3. The operation is repeated until about 30 GVs are collected (*see* **Note 11**).

4. The GV containing the transplanted nuclei are then disrupted by brief vortexing (5 s), and centrifugated for 3 min at $12,000 \times g$.

5. The supernatant is discarded and the pellets containing the nuclei and GV membrane is washed twice in 1 ml of non denaturing washing buffer by vortexing 5 s and centrifugating for 3 min at $12,000 \times g$.

6. The obtained pellets correspond mostly to transplanted chromatin components since the washing process gets rid of most of the GV proteins that are not tightly bound to the transplanted nuclei.

7. The pellets can be frozen or directly processed for WB analysis following a standard protocol.

3.2.3 Collection of Samples for ChIP Analysis

ChIP analysis from nuclei transplanted to *Xenopus* oocyte is also possible, despite the relatively low amount of nuclei used (*see* **Note 12**).

1. Oocytes are washed in MBS 1× without BSA prior to fixation.

2. Batch of oocytes (typically 56–84 oocytes) are transferred to a 15 ml falcon tube.

3. 8 ml of MBS 1 % formaldehyde are added to the oocytes and the samples are fixed for 10 min at room temperature with gentle agitation.

4. The fixation is stopped by the addition of 4 ml of MBS 1× with 300 mM glycine.

5. The oocytes are then washed three times in 10 ml MBS 1×.

6. Groups of seven fixed oocytes are transferred to Eppendorf tubes in which MBS is replaced by 280 μl of homogenization buffer supplemented with 0.3 % SDS, proteases inhibitors, and 1 mM DTT.

7. Oocytes are disrupted by pipetting up and down using a micro-pipette (*see* **Note 13**).

8. Sonication is carried out in two cycles of 7 min (with 30 s on/off cycles, higher power).

9. After sonication, the SDS in the sample is diluted by adding 600 μl of homogenization buffer supplemented with proteases inhibitors and 1 mM DTT.

10. The sonicated chromatin is then centrifugated at $12,000 \times g$ for 10 min at $4°$.

11. The supernatant (including the lipid layer) is then transferred to a new tube.

12. This chromatin can then be used in a standard ChIP protocol.

3.3   Live Imaging of Transplanted Nuclei

Nuclear transfer to *Xenopus* oocyte offers the possibility of performing live imaging of transplanted nuclei. This is done by isolating the oocyte germinal vesicle in nonaqueous medium. Under the right condition these germinal vesicles are able to sustain transcription in transplanted nuclei for up to 18 h [3, 10].

3.3.1   Extraction of GV in Oil

Isolation is carried out at room temperature.

1. To isolate the GVs, group of ten oocytes are transferred to nonaqueous medium (mineral oil) using a glass pipette (*see* **Note 14**).

2. Any excess of aqueous medium surrounding the oocytes needs to be carefully removed using a micropipette, first with a 200 μl tip and then with a 20 μl tip (Fig. 4) (*see* **Note 15**).

3. The GVs is then taken out of the oocyte by opening the animal pole of the oocyte with forceps (Fig. 4) (*see* **Note 16**).

3.3.2   Nuclear Transfer and Live Imaging

1. The transplantation is carried out immediately after GV isolation (*see* Subheadings 3.1.3 and 3.1.4).

2. For imaging the transplanted nuclei, oil-GV are transferred to the shallow well of imaging slides and gently flattened under a coverslip (Fig. 4) (*see* **Note 17**).

3. Confocal microscope analysis can then be carried out on an upright confocal microscope at a temperature of 18–20 °C, using classical live imaging condition.

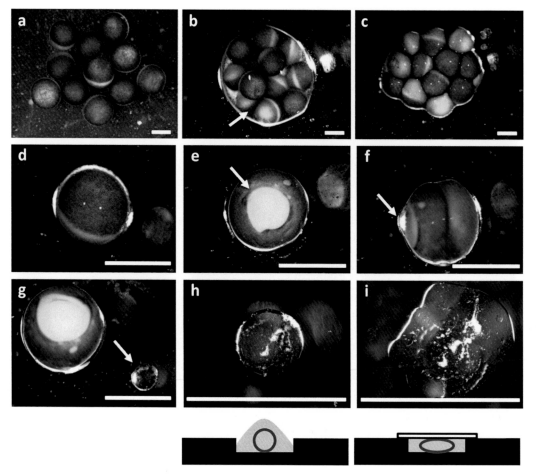

Fig. 4 Isolation of germinal vesicle in mineral oil for live imaging (**a**) Sorted stage V–VI defolliculated oocytes in
1× MBS. (**b**) Oocytes transferred to mineral oil. The MBS surrounding the oocyte is still visible (*arrow*). (**c**) After
the MBS has been sucked away with a micropipette the oocytes are left in mineral oil with a minimal amount
of aqueous medium surrounding them. (**d**) Individual oocytes can then be separated (**e**) The membrane at the
animal pole is split open with dissecting forceps so that the underlying germinal vesicle is revealed (*arrow*)
(**f**) Dissected oocyte shown sideway to illustrate the position of the germinal vesicle protruding from the oocyte
yolk at the dissected animal pole (*arrow*). (**g**) The germinal vesicle is pushed away free of yolk from the oocyte
(**h**) The germinal vesicle (*close up image*) is then transferred with the surrounding mineral oil to a slide with
shallow well (*diagram, bottom left*) (**i**) The germinal vesicle is then gently flattened under a coverslip

4 Notes

1. Injection needles can be pulled in a Flaming/Brown micropipette puller (Sutter instrument model P87 or p-7) using the
   following parameters: *heat*: set at the temperature established
   by a ramp test, *pull*: 100, *velocity*: 100, *time*: 10.

2. Protein expression from injected mRNA is more efficient when
   the mRNA is polyA-tailed and contains modified GTP
   (m7G(5′)ppp(5′)G).

3. In order to remove the follicular cells that surround the oocytes and prevent microinjection, ovaries are treated with an enzymatic cocktail that loosen intercellular connections.

4. The quantity of mRNA to be injected depends on the protein of interest. Typically, 2–27 ng of mRNA is injected in a volume of 2.3–40 nl.

5. The nuclei used for nuclear transfer can be prepared using various reagents such as digitonin, lysolecithin, or streptolysin O (SLO). I describe here the use of SLO as it is commonly used. As compared to digitonin and lysolecithin, SLO is a milder reagent as it permeabilizes only the cytoplasmic membrane but leaves the nuclear membrane functional.

6. This permeabilization conditions are adapted for common cells, such as MEF or ES. However, for other types of cells it might be necessary to alter the ratio of cells to SLO, as the activity of this permeabilization reagent depends on the amount and composition of plasma membrane.

7. It is possible to store such nuclei at −80 °C. If doing so, it is important to avoid several freeze/thaw cycle as this will damage the nuclei and release DNA making the nuclei suspension difficult to handle for transplantation.

8. The difficulty of transplanting nuclei to *Xenopus* oocyte lays in the fact that it is not possible to visualize the target germinal vesicle. Fully grown *X. Laevis* oocytes are about 1.2 mm in diameter and house a germinal vesicle 400 μm in diameter. The GV is located at the animal pole of the oocyte, about 100 μm beneath the oocyte membrane, therefore marking the needle beforehand will help.

9. It is important to note that very often the nuclei suspension, when sucked into the needle, does not remain homogeneous. Indeed the nuclei tend to become more concentrated at the oil/nuclei solution interface. For that reason it is crucial to refill the needle frequently and to then immediately transplant the nuclei to oocyte. This operating mode ensures that the quantity of nuclei delivered to the GV is constant.

10. To reduce mRNA loss, the entire process needs to be carried out rapidly (less than a minute between dissection and transfer to dry ice).

11. Thirty GVs correspond to roughly 7,500 transplanted nuclei, yielding enough material for detection using antibodies against modified histones, for example.

12. It is likely that the large amount of proteins and RNAs accumulated in *Xenopus* oocytes acts as a carrier and helps the process. Additionally, one has to realize that the presence of *Xenopus* follicular cells left over after liberase treatment can contribute to a large fraction of the chromatin obtained from transplanted oocytes. This parameter is important to consider especially if

ChIP-seq analysis is envisaged, since it will affect the percentage of immunoprecipitated material that will correspond to the transplanted mouse nuclei or to the *Xenopus* recipient.

13. It is very important at this stage to thoroughly homogenize the oocytes.

14. Isolation in oil prevents any leakage of nuclear components into the culture medium.

15. The best results are achieved by aspirating the excess of MBS with the pipette tip positioned in-between the oocytes. It is very important to get rid of the aqueous medium before proceeding to GV isolation since any left over MBS is going to penetrate the GV and dilute its content.

16. At this stage it is essential to remove the yolk granules that tend to stick to the surface of the germinal vesicle. This is best obtained by "opening" the oocyte quite widely, let it stand for 1–2 min, and then proceed to gently push the GV away from the oocyte cytoplasm.

17. It is important not to squash the GVs as the transcriptional activity is lost upon disruption of the GV integrity.

Acknowledgements

This work was supported by grants from the Wellcome Trust and the Medical Research Council, UK.

References

1. Jullien J, Pasque V, Halley-Stott RP et al (2011) Mechanisms of nuclear reprogramming by eggs and oocytes: a deterministic process? Nat Rev Mol Cell Biol 12:453–459

2. Jullien J, Astrand C, Szenker E et al (2012) HIRA dependent H3.3 deposition is required for transcriptional reprogramming following nuclear transfer to Xenopus oocytes. Epigenetics Chromatin 5:17

3. Jullien J, Astrand C, Halley-Stott RP et al (2010) Characterization of somatic cell nuclear reprogramming by oocytes in which a linker histone is required for pluripotency gene reactivation. Proc Natl Acad Sci U S A 107:5483–5488

4. Miyamoto K, Pasque V, Jullien J et al (2011) Nuclear actin polymerization is required for transcriptional reprogramming of Oct4 by oocytes. Genes Dev 25:946–958

5. Barreto G, Schäfer A, Marhold J et al (2007) Gadd45a promotes epigenetic gene activation by repairmediated DNA demethylation. Nature 445:671–675

6. Pasque V, Gillich A, Garrett N et al (2011) Histone variant macroH2A confers resistance to nuclear reprogramming. EMBO J 30: 2373–2387

7. Byrne JA, Simonsson S, Western PS et al (2003) Nuclei of adult mammalian somatic cells are directly reprogrammed to oct-4 stem cell gene expression by amphibian oocytes. Curr Biol 13:1206–1213

8. Halley-Stott RP, Pasque V, Astrand C et al (2010) Mammalian nuclear transplantation to Germinal Vesicle stage Xenopus oocytes—a method for quantitative transcriptional reprogramming. Methods 51:56–65

9. Murata K, Kouzarides T, Bannister AJ et al (2010) Histone H3 lysine 4 methylation is associated with the transcriptional reprogramming efficiency of somatic nuclei by oocytes. Epigenetics Chromatin 3:4

10. Lund E, Paine PL (1990) Nonaqueous isolation of transcriptionally active nuclei from Xenopus oocytes. Methods Enzymol 181:36–43

Chapter 7

Assessing the Quality of Donor Cells: Karyotyping Methods

Amélie Bonnet-Garnier, Anne-Clémence Veillard, Bertrand Bed'Hom, Hélène Hayes, and Janice Britton-Davidian

Abstract

Somatic cell nuclear transfer (SCNT) has a low success rate that rarely exceeds 5 %. Moreover, SCNT requires highly technical skills and may be influenced by the biological material used (oocyte and donor cell quality). Hence, it is crucial to check the normality of the donor cell's karyotype. Numerical and structural chromosome abnormalities are detected by cytogenetic analysis at minimum using G-banding to identify the chromosomes. Here, we describe the classical protocols that are needed to perform complete cytogenetic analyses, i.e., G-banding to identify chromosome aberrations, followed by Fluorescent In Situ Hybridization (FISH) of specific probes for a more sensitive detection and precise identification of the rearrangement.

Key words Karyotype, Embryonic stem cells, G-banding, FISH

1 Introduction

1.1 Why Is It Necessary to Karyotype the Donor Cells?

Somatic cell nuclear transfer (SCNT) is a very complicated technique that requires specific skills. Since the efficiency of SCNT is extremely low (from 3 to 5 % of reconstructed or cloned embryos develop to term [1]), it is of real importance to control as much as possible the parameters that can influence the outcome of SCNT. Some parameters are difficult to normalize from one experiment to another (for example oocyte quality or recipient animal), but the stability and normality of the donor cell lines can be monitored by karyotype analysis.

Different kinds of donor cells, according to the species cloned, are classically used in our laboratory: (1) cumulus cells (belonging to the *corona radiata* that surrounds the oocyte) and in this case, no karyotype is needed, or (2) cells in culture. Indeed, it has been reported that primary cell lines or mouse embryonic stem cell lines [2, 3] that have undergone a high number of passages or prolonged cell culture, have a significant percentage of chromosome

Nathalie Beaujean et al. (eds.), *Nuclear Reprogramming: Methods and Protocols*, Methods in Molecular Biology, vol. 1222, DOI 10.1007/978-1-4939-1594-1_7, © Springer Science+Business Media New York 2015

abnormalities (gain or loss of a chromosome or more complex rearrangements). Thus, it is necessary to assess the stability of the karyotype at different time points during cell culture to avoid failure during development of the cloned embryo that originates from the donor cell nucleus.

In this chapter, we will describe detailed protocols for assessing chromosome normality or abnormality (structural changes or aneuploidies, i.e., monosomy or trisomy) of donor cells with cytogenetic methods. The chapter will describe (1) methods to obtain high-quality metaphase spreads from different cells in culture, (2) several classical cytogenetic procedures (i.e., C-banding adapted from Sumner [4] and G-banding adapted from Seabright [5]) that allow chromosome numbering and karyotype construction and (3) a molecular cytogenetic technique (hybridization of specific probes revealed with fluorescence also called Fluorescent In Situ Hybridization—FISH) that provides an accurate means of detecting and/or identifying chromosome abnormalities.

2 Materials

Solutions are sterilized and filtered using MF-Millipore membrane filters (0.22 μm). Unless otherwise specified, solutions and reagents are stored at room temperature (RT).

2.1 Cell Synchronization and Harvesting

1. Donor cells: bovine ear fibroblasts (BEFs), mouse embryonic stem cells (mESCs) or mouse epiblast stem cells (mEpiSCs) in culture (see **Note 1**).

2. Dulbecco's Phosphate-Buffered Saline (DPBS) 1× sterile without Ca^{2+} and Mg^{2+}.

3. Culture medium adapted to each cell type 1× supplemented with glutamine, specific cell growth factors, and fetal calf serum (see **Note 1**), stored at 4 °C.

4. Thymidine 1× solution (10 mg/mL): dissolve 100 mg of thymidine in 10 mL of culture medium. Filter and sterilize before storing at 4 °C up to 1 week.

5. FdU 10× solution (1 mg/mL): dissolve 10 mg of 5-fluoro-2′-deoxyuridine (FdU) in 10 mL of DPBS. Store 1 mL aliquots of the 10× solution in 1.5 mL tubes at −20 °C. The working solution (100 μg/mL) is prepared from 1 mL of stock solution diluted with 9 mL of DPBS. After filtration and sterilization, store working solution (1×) in 500 μL aliquots at −20 °C.

6. BrdU 1× solution (1 mg/mL): dissolve 10 mg of 5-bromo-2′deoxyurine (BrdU) in 10 mL of DPBS. Filter and sterilize before storing 500 μL aliquots at −20 °C.

7. Colchicine stock solution (10×, 40 µg/mL): dissolve 0.4 mg of colchicine powder in 10 mL of DPBS. Store 1 mL aliquots of 10× solution in 1.5 mL tubes at –20 °C. The working solution (1×) is prepared from 1 mL of stock solution diluted in 9 mL of DPBS. Aliquots of 500 µL of working solution (1×) can be kept up to 1 year at 4 °C or stored at –20 °C.

8. Trypsin-EDTA 1× solution prepared with 0.5 g/L of trypsin and 0.2 g/L of EDTA. Store 10 mL aliquots at –20 °C.

9. Hank's Balanced Salt Solution without Ca^{2+} and Mg^{2+} (HBSS⁻).

10. Hank's Balanced Salt Solution with Ca^{2+} and Mg^{2+} (HBSS⁺).

11. Potassium chloride solution (KCl) 10× (56 g/L): dissolve 56 mg of KCl in 10 mL of ultrapure water (produced by a Milli-Q Water System (Millipore) with a resistivity of 18.2 MΩ.cm at 25 °C).

12. Sterile Fetal Calf Serum (FCS) or Newborn Calf Serum (NCS).

13. Hypotonic solution (*see* **Note 2**): the working solution is a mix (in various proportions—*see* Table 1) of deionized water (sterile, for injection), FCS, and KCl 10× solution. Warm the hypotonic solution in a water bath at the desired temperature (*see* Table 1 for details). Prepare a fresh solution just before use.

14. Fixative: mix 3 volumes of absolute ethanol with 1 volume of acetic acid. Prepare the solution extemporaneously and keep cold at 4 °C until use.

Table 1
Hypotonic solutions with different proportions of FCS and KCl

Proportion FCS:KCl	FCS (mL)	KCl 10× (mL)	Water (mL)	Incubation time (min) at 37 °C
100:0	5	0	25	15
90:10	4.5	0.3	25.2	14
80:20	4	0.6	25.4	13
70:30	3.5	0.9	25.6	12
60:40	3	1.2	25.8	11
50:50	2.5	1.5	26	10
40:60	2	1.8	26.2	9
30:70	1.5	2.1	26.4	8
20:80	1	2.4	26.6	7
10:90	0.5	2.7	26.8	6
0:100	0	3	27	5

15. 15 mL sterile conical centrifuge tubes and 25 mL sterile plastic pots.

16. Graduated plastic pipettes (sterile, single package) of 2, 5, 10, and 25 mL.

17. Glass Pasteur pipettes (length 230 mm, cotton plugged). Sterilize the pipettes in an aluminum container using a dry oven (4 h at 180 °C).

18. Plastic sterile Petri dishes (60 mm diameter) or plastic sterile flasks (75 cm^2).

19. Water bath.

20. Centrifuge (up to $583 \times g$ for 15 mL plastic tubes).

21. Inverted microscope with phase contrast and a 10× magnification objective to monitor cell morphology and mitotic index.

2.2 Metaphase Spreading and Conventional Staining

1. Microscope slides: superfrost ultra plus, precleaned, ready-to-use, round edge slides.

2. Glass Pasteur pipettes (length 230 mm).

3. Sørensen buffer solution at pH 6.8: Prepare a solution A (9.08 g of KH_2PO_4 in 1 L of ultrapure water) and a solution B (23.88 g of $Na_2HPO_4 \cdot 12H_2O$ in 1 L of ultra-pure water), then mix 508 mL of solution A with 492 mL of solution B to obtain a buffer solution at pH 6.8.

4. Giemsa stain solution (4 %): mix 4 mL of Giemsa and 4 mL of Sørensen buffer in 92 mL of distilled water, then filter.

5. Several coplin (porcelain or glass) jars (100 mL) for staining.

6. Phase-contrast microscope equipped with a 10× magnification objective, a 40× or 63× magnification oil objective and a 100× magnification oil objective.

2.3 C-Banding

1. 20× SSC stock solution: 17.53 g of NaCl, 8.82 g of citrate trisodium in up to 100 mL of ultrapure water and adjust with hydrochloric acid (HCl 12 N) to pH 5.8 (for dilution with formamide) or pH 6.3 (for dilution with distilled or deionized water). Filter and keep at room temperature up to several months. The 2× SSC working solution is obtained by a 10× dilution of the 20× SSC stock solution.

2. Baryta (barium hydroxide, $Ba(OH)_2$) saturated solution: dissolve 4.6 g $Ba(OH)_2$ in 100 mL of distilled water by continuous stirring during 1 h, then leave to stand overnight before use.

3. Hydrochloric acid (HCl) solution 0.2 N: mix 20 mL of HCl 2 N solution with 80 mL of distilled water.

4. Two water baths that can reach a temperature of 100 °C.

2.4 G-Banding

1. Saline solution (0.9 %): dissolve 0.9 g of NaCl in 100 mL of distilled water.

2. Disodium phosphate buffer (0.25 M): dissolve 17.8 g of disodium hydrogenophosphate dodecahydrate ($Na_2HPO_4 \cdot 12H_2O$) into 250 mL of distilled water.

3. Trypsin solution (0.25 %): dissolve 250 mg of trypsin in 75 mL of saline solution and add distilled water up to 100 mL. Adjust the pH at 7.0 with Na_2HPO_4 and chill the solution to 11–12 °C before use (the solution must be prepared extemporaneously, *see* **Note 3**).

4. Phosphate-Buffered Saline (PBS) 1×: dissolve one ready-to-use tablet in 100 mL of deionized water and autoclave to sterilize.

5. Citric acid solution (1 M): dissolve 5.25 g of citric acid in 250 mL of distilled water.

6. Giemsa for G-banding (3 %): mix with stirring 90 mL of water, 3 mL of Giemsa, 3 mL of methanol (under fume hood) and 3 mL of citric acid solution, then adjust to pH 6.8 with the disodium phosphate buffer (about 1 mL).

7. A dry oven or a hot plate that can reach a temperature of 100 °C.

2.5 FISH

1. 2× SSC working solution (pH 7.0): obtained by diluting to 10× the 20× SSC solution at pH 6.3.

2. Two series of absolute ethanol 100 %, 75 % and 50 % in coplin jars: one stored at −20 °C and the other at room temperature.

3. RNase stock solution (20 mg/mL): dissolve 200 mg of RNase powder (type IA) in 10 mL of water, boil during 10 min and store at −20 °C in 1 mL aliquots.

4. RNase working solution (0.2 mg/mL): dilute stock solution to 1/100 in 2× SSC (pH 6.3).

5. Pepsin stock solution (10 mg/mL): dissolve 25 mg of pepsin in 25 mL DPBS (without Ca^{++} and Mg^{++}) and store at −20 °C in 500 µL aliquots.

6. Hydrochloric acid solution (HCl) 0.01 N: dilute a 1 N HCl solution to 1/100 in deionized water.

7. Pepsin working solution (20 µg/mL) pH 3.0: dilute 200 µL of pepsin stock solution in 100 mL of 0.01 N HCl solution and warm to 37 °C.

8. $MgCl_2$ working solution (50 mM): dissolve 101.5 mg of $MgCl_2 \cdot 6H_2O$ in 100 mL of distilled water.

9. PBS–$MgCl_2$ solution: Mix 50 mL of 1× PBS and 50 mL of $MgCl_2$ working solution.

10. Formamide solution 70 %: mix 70 mL of pure deionized formamide with 10 mL of 20× SSC (pH 5.8 previously adjusted with HCl 12 N) and add deionized water up to 100 mL.

11. Humid dark chamber at 37 °C.

12. Rubber cement.

13. Hot plate to warm the slides up to 100 °C.

14. Stirring water bath (up to 100 °C) placed under a fume hood.

15. Plastic and glass cover slips.

16. Hybridization buffer and chromosome painting probes. Store at −20 °C until use (*see* **Note 4**).

17. Formamide solution 50 %: mix 100 mL of pure formamide with 20 mL of 20× SSC (pH 5.8) and add deionized water up to 200 mL.

18. PBS–BSA–0.01 % Tween 20 solution (PBT): mix 100 mL of (sterile) PBS and 400 μL Bovine Serum Albumin (purchased as a 30 %—liquid solution) and 100 μL Tween 20 (10 %—liquid solution). Store at 4 °C up to 4 weeks.

19. 4× SSC—0.02 % Tween 20 solution: mix 20 mL of 20× SSC (pH 6.3) with 79.8 mL of deionized water and add 200 μL of Tween 20. Store at 4 °C up to 4 weeks.

20. Primary and secondary antibodies if required depending on the probe labeling (*see* **Note 5**); standard dilution 1/200 in PBT.

21. Commercial anti-fading solution: Vectashield.

22. PPD 11 (*p*-PhenyleneDiamine) at pH 11.0: dissolve 100 mg of PPD (powder) in 10 mL of PBS at 37 °C with stirring protected from light. Add 90 mL of glycerol and homogenize during 2 h by continuous stirring. Adjust to pH 11 by adding NaOH (1 M) very slowly for the pH to stabilize. Store in 100 μL aliquots protected from light at −20 °C.

23. 4′, 6-diamidino-2-phenylindole dihydrochloride (DAPI): Prepare a stock solution at 100 μg/mL (dissolve 1 mg of powder in 10 mL of deionized water and store in 100 μL aliquots at −20 °C) and dilute 2 μL of this stock solution in 198 μL of Vectashield.

24. Propidium Iodide (PI): Prepare a stock solution at 100 μg/mL (dissolve 1 mg of propidium iodide powder in 10 mL of deionized water and store in 100 μL aliquots at −20 °C) and dilute 4 μL of this stock solution in 600 μL of PBT and use PPD 11 as mounting medium.

25. Microscope equipped with a 10× magnification phase contrast objective, 63× magnification (oil) and 100× magnification (oil) objectives (fluorescent Plan Apochromat), a motorized filter wheel, a camera, and an informatics system (hardware and software) dedicated to cytogenetic/FISH imaging (*see* **Note 6**).

3   Methods

3.1   Cell Culture Dilution or Passage

To obtain mitotic and more specifically metaphase cells during harvest, it is necessary that the cell culture is in a proliferative phase. Thus, the first step consists in splitting confluent cells to three dishes or flasks (one-third dilution).

1. Warm appropriate culture medium, HBSS- and trypsin/EDTA or collagenase II at 37 °C before use.

2. The cells are cultured in dishes or flasks in appropriate culture medium (according to cell type, *see* **Note 1**), using an incubator at 37 °C in a humid atmosphere with 5 % CO_2.

3. Remove the culture medium from the dishes (or flasks) by aspiration with a sterile glass Pasteur pipette.

4. Then add 0.1 mL of HBSS- for 5 min at 37 °C to eliminate excess medium, remove HBSS- and repeat this step once.

5. Incubate with 50 μL of the trypsin/EDTA solution per cm^2 for 1–5 min at 37 °C to resuspend the cells (except for EpiSCs where 40 μl of preparation of collagenase II solution is discribed elsewhere (*see* **Note 1**), are added per cm^2 for 1–2 min).

6. When the cells are detached, inactivate trypsin/collagenase by adding appropriate culture medium supplemented with FCS (10 %) (for EpiSCs, withdraw the collagenase, rinse carefully with DPBS, add culture medium and scrape the Petri dish with a 1 mL tip) and transfer the cell suspension into 10 mL tubes and centrifuge at $180 \times g$ for 5–10 min.

7. Discard the supernatant and resuspend the pellet with a glass Pasteur pipette in 1 mL of culture medium. In the case of EpiSCs do not resuspend the pellet too much to preserve clumps of EpiSCs.

8. Distribute the cells into three or four dishes (or flasks) to maintain the cells in division (they should not reach confluence after 48 h).

3.2   (Double) Synchronization with Thymidine (See Note 7 for an Alternative Method)

Synchronization of cells is used to enrich the number of metaphase cells in the cell culture at harvest. To synchronize cells, a large amount of thymidine (5 mM) is added to the medium which induces an arrest of cell replication. A double synchronization consists in repeating the blocking procedure after removing thymidine and letting the cells divide once. To remove thymidine, eliminate the culture medium supplemented with thymidine, rinse several times the cell layer and add standard cell culture medium, then a majority of the cells will restart their cell cycle in S phase at the same time.

Day 1

1. Divide the cell cultures as described above (Subheading 3.1).

2. 4 h later (when cells are attached to the surface of the flask), add 200 μL of the thymidine solution per mL of medium.

Day 2

3. Fourteen hours (H18 from Day 1) later, remove the medium and rinse twice with HBSS⁺ as described in Subheading 3.1, **step 4**. Add fresh medium (0.1 mL of culture medium per cm²).

4. Ten hours later (H28 from Day 1), add 200 μL of the thymidine solution per mL of medium.

Day 3

5. Fourteen hours (H42 from Day 1) later, remove the medium and rinse twice with HBSS⁺ as described in Subheading 3.1, **step 4**. Add fresh medium (0.1 mL of culture medium per cm²).

6. Harvest the cells between 6 and 10 h later (H48–H52 from Day 1) when the wave of mitosis begins. Observe dishes every 15 min to determine when the mitotic index of the cell culture has reached a peak (which can be calculated by dividing the number of cells in mitosis by the number of cells in telophase).

3.3 Harvesting

Colchicine (demecolcine or colcemid) is a drug that inhibits the formation of the mitotic spindle and prevents the division of chromosomes at anaphase. The chromosomes remain in metaphase and continue to condense until the drug is removed. The longer the colchicine acts the higher will be the number of cells in metaphase but the chromosomes will also become more condensed thus shorter, which can make accurate karyotyping difficult. Performing synchronization or double synchronization of the cell culture helps to circumvent this problem.

1. Add 10 μL of colchicine per mL of culture medium in the culture dish for 30–45 min at 37 °C in the CO_2 incubator.

2. Meanwhile, warm the hypotonic solution in a water bath at 37 °C, prepare the fixative and chill at 4 °C.

3. Transfer the culture medium in a 15 mL conical centrifuge tube.

4. Rinse with 2 mL of HBSS⁻ that will be transferred into the same 15 mL tube.

5. Incubate the adherent cells with 0.5 mL of trypsin/EDTA solution for 2–5 min maximum at 37 °C in the incubator.

6. Inactivate with 2 mL of medium supplemented with 10 % of FCS and transfer the detached cells in the same 15 mL tube.

7. Centrifuge for 5 min at 1,500 rpm.

8. Discard the supernatant.

9. Resuspend gently with a glass Pasteur pipette (avoid bubbles) in a few drops of hypotonic solution until the pellet is completely homogenized and fill up to 10 mL with the hypotonic solution.

10. Incubate in the water bath at 37 °C for 5–15 min according to the hypotonic treatment used (*see* Table 1).

11. At the end of the incubation, remove the tube from the water bath, add 1 mL of cold fixative and gently invert the tube.

12. Wait 5 min at room temperature and centrifuge 10 min at $405 \times g$.

13. Discard the supernatant and resuspend the pellet gently in a small volume (500 µL) of ice-cold fixative with a glass Pasteur pipette and fill up to 10 mL when the suspension is homogenized.

14. Store at 4 °C for 20 min.

15. Centrifuge for 5 min at $405 \times g$ and discard the supernatant.

16. Resuspend into 8–10 mL cold fixative and store overnight at 4 °C.

Chromosome spreading can be done the next day or the fixed cells can be stored at –20°C in 1 mL aliquots.

3.4 Metaphase Spreading

1. Centrifuge the tubes containing the fixed cells for 10 min at $583 \times g$.

2. Remove the supernatant by aspiration taking care to preserve the pellet and resuspend the pellet gently (to avoid bubbles) in 500 µL to 1 mL fresh fixative (approximately fourfold the volume of the pellet) with a glass Pasteur pipette.

3. Let one drop (~20 µL) of the cell suspension fall on a dry slide from a height of ±30 cm and blow gently on the slide to improve spreading.

4. Allow to dry under a fume hood and then examine the density of the metaphase spreads under a microscope with phase contrast to adjust the dilution of the cell suspension with fixative; continue metaphase spreading.

3.5 Conventional Stain

1. Place slides in a coplin jar containing the 4 % Giemsa solution and incubate 5–10 min.

2. Rinse with tap water then distilled water.

3. Dry under a fume hood.

4. Observe the slides using a bright field microscope with the 10× objective and screen for well-spread and easy to count metaphases (metaphases should be clearly separated so that chromosomes do not overlap but are not too dispersed).

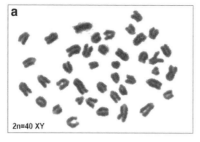
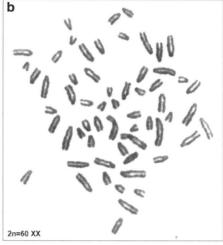

Fig. 1 Metaphase cells with satisfactory chromosome spreading prepared from a mouse embryonic stem cell culture (**a**) and a bovine fibroblast primary cell culture (**b**). The chromosomes are stained with Giemsa alone

5. Capture images at 40× and 100× with a dedicated software for cytogenetic analysis (*see* **Note 6**). Figure 1a, b show examples of mouse and bovine chromosomes stained with Giemsa alone, which allows counting the diploid chromosome number count and distinguishing between metacentric and acrocentric chromosomes.

3.6 C-Banding

The C banding technique is used to selectively reveal constitutive heterochromatine. It stains mainly the centromeric and the pericentromeric regions and is useful to detect loss of centromeres or chromosome fusions (such as Robertsonian or tandem translocations). With this technique it is easy to identify chromosome morphology and to distinguish acrocentric from metacentric or submetacentric chromosomes.

Between each incubation, slides are rinsed with tap water.
Warm one water bath at 50°C and one at 60°C.

1. Immerse the slides in 0.2 N HCl for 1 h at room temperature, then rinse.

2. Incubate slides in the barium solution at 50 °C for 1–3 min, plunge briefly the slides in 0.2 N HCl to remove any deposit of barium and rinse.

3. Wash slides in 2× SSC at 60 °C for 20 min.

4. Stain slides for 10 min in a solution of 4 % Giemsa and rinse them with running tap water. Let the slides dry before observation.

3.7 G-Banding Using Trypsin Treatment and Revealed by Giemsa (GTG)

G-bands correspond to A-T rich/gene-poor regions of the chromatin, which are insensitive to treatment with a proteolytic enzyme (trypsin) and thus can be counterstained with Giemsa. This technique results in an alternating pattern of dark and light bands along the metaphase chromosome, which is chromosome specific. This specific banding pattern allows chromosome pairing (karyotyping) and identification of chromosome abnormalities or rearrangements according to the standard karyotype established for each species.

Day 0

Age freshly spread slides in an oven at 40 °C overnight. This aging procedure improves the standardization of the trypsin treatment.

Day 1

1. Warm a 2× SSC solution at 60 °C in a 100 mL coplin jar.

2. Immerse the slides in 2× SSC at 60 °C for 1 h (prepare the trypsin solution during this time and cool it to 11 °C).

3. Rinse the slides under tap water then distilled water and let them dry.

4. Prepare three coplin jars (CJ): CJ1 with NaCl solution, CJ2 with trypsin solution and CJ3 with PBS.

5. Humidify the slides in the NaCl solution (CJ1).

6. Incubate slides in the trypsin solution at 11 °C (CJ2 chilled on ice) for 1 min (at least two different incubation times (±15 s) should be tested).

7. Rinse immediately in PBS (CJ3).

8. Rinse with distilled water, then let air dry.

9. Counterstain the slides with Giemsa 3 % for 10 min at room temperature.

10. Examine under the microscope with the objective 40× (at minimum) to evaluate if the trypsin treatment is satisfactory. If chromosomes are not sufficiently digested (dark without banding) or too digested (puffy and light), adjust trypsin incubation times (increase or reduce) until the optimal time is reached (the same day with the same slides).

We recommend that at least ten high-quality G-banded chromosome metaphases are analyzed to confirm a normal karyotype or to correctly identify chromosome abnormality(ies) of a cell culture.

Figure 2a–c show a good-quality G-banded metaphase and two different karyotypes from the same cell line (n°3.83) of mouse epiblast stem cells. This cell culture shows a trisomy of mouse chromosome 11 (MMU11) in a mosaic state, with 23 % of the metaphases counted showing a normal karyotype with $2n=40$, while 50 % and 27 % of the metaphases counted had respectively a third MMU11 either free (Fig. 2b, $2n=41$) or translocated onto another chromosome 11 (Fig. 2c, $2n=40$).

This example shows that counting the chromosomes with Giemsa staining only (and even more so by an inexperienced

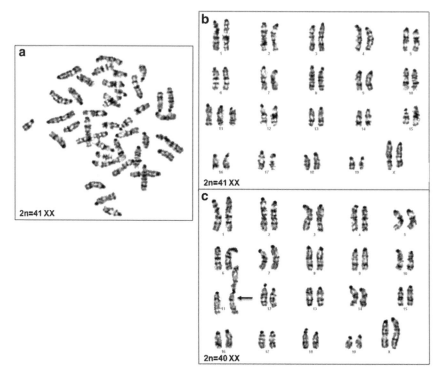

Fig. 2 Image (**a**) shows a metaphase with satisfactory chromosome spreading and G-banded chromosomes from an EpiSC line (n° 3.83). (**b** and **c**) Show karyotypes with two different kinds of chromosome 11 trisomy: (**b**) the three chromosomes are independent and (**c**) two chromosomes 11 are fused (*arrow*) by their centromere (Robertsonian fusion)

person) may mislead one to believe that the cell culture is normal because the diploid number is correct although the morphology of some chromosomes is abnormal. Indeed, reciprocal translocations, inversions (for example) or trisomies with a Robertsonian translocation do not change the diploid number [6].

3.8 FISH (Fluorescent In Situ Hybridization)

This technique allows fluorescent visualization of a specific hybrid between a nucleotide sequence (RNA or DNA probe) and a biological sample (chromosome spread, nucleus, or tissue section). It relies on the formation of DNA/DNA or DNA/RNA hybrids that is possible only if the sequences share a sufficient level of sequence similarity and if the DNA double strands are denatured. Visualization of the hybrids depends on the use of a probe labeled either directly with a flurochrome or indirectly with a conjugate that will be detected with a fluorescent antibody.

*3.8.1 Slide Pretreatment (See **Note 8**)*

1. Warm a water bath at 38 °C.

2. Incubate the slides 1 h in a coplin jar containing 100 mL of RNase working solution.

3. Rinse at room temperature 10 min in another 100 mL coplin jar with 2× SSC.

4. In a 38 °C water bath, incubate the slides 10 min in the pepsin working solution.

5. Stop the reaction with a cold bath of PBS-MgCl₂ solution.

6. Dehydrate the slides through a series of increased alcohol concentrations in coplin jars (50, 75, and finally 100 %, 10 min each) and allow the slides to dry completely (after this step, slides can be frozen at −20 °C in a box with a desiccant and stored several months until use).

3.8.2 Hybridization and Signal Detection

1. Probe denaturation step: Prewarm the probes (whole-chromosome painting probes) and the hybridization buffer at 37 °C for 5–10 min. Vortex and spin down briefly (or follow manufacturer's recommendations). Mix (in a 500 μL micro-tube) an adequate amount of probe (*see* manufacturer's recom-mendations) and Hybridization buffer solutions to a final volume of 15–20 μL (according to the size of the cover slip: 18 × 18 or 22 × 22 mm). Warm the probe mixture to 75 °C for 10 min. Centrifuge and keep the probe mixture at 37 °C in another water bath for a 30–45 min prehybridization step (which consists in suppressing repeated sequences by hybrid-ization with the Cot-1 DNA included in the commercial probe solution). Synchronize probe and slide denaturation steps.

2. Slide denaturation: Prewarm a glass jar containing the 2× SSC/70 % formamide solution in a water bath at 75 °C. Check the temperature (72 °C inside the jar) and incubate each slide (one by one) during exactly 2 min at exactly 72 °C. Immediately dehydrate each slide in a glacial series of increased ethanol con-centrations in coplin jars (50, 75, and finally 100 %, 2 min each) to avoid DNA rehybridization (coplin jars with the etha-nol solutions are kept at −20 °C until use). Let the slides dry for a few minutes.

3. Hybridization step: Load 15 μL of probe mixture on each dry "denaturated" slide. Cover the zone of each slide containing the metaphase spreads (previously delimited with a diamond-head pen) with a 18 × 18 glass cover slip. Seal with rubber cement. Set the slide in an airtight box (humidified with a for-mamide 50 %/2×SSC solution) in a 37 °C oven during 18 (overnight) to 48 h (longer hybridization periods will increase background).

4. Washing step: After hybridization, remove the rubber cement and the cover slip carefully. Wash the slides at 42 °C (warmed in a water bath) 3 min successively in each of two coplin jars containing 50 % formamide/2× SSC and then two others with only 2× SSC. These washes will remove probe in excess and reduce background.

If probes are already coupled with a fluorescent dye go directly to the DNA staining step.

5. Detection of digoxigenin-coupled or biotin-coupled probes (optional): Place the slides 10 min in a coplin jar containing PBT solution at room temperature to saturate nonspecific sites on the slides. Remove slides from the coplin jar and load 60 μL of anti-digoxigenin or anti-biotin antibody (produced in mouse, sheep, or rabbit; dilution according to manufacturer's recommendation) on each slide and cover with a glass cover slip (24 × 60 mm). Incubate 45 min to 1 h at 37 °C in a humid container (do not let the slides dry). Then wash with PBS-Tween 20 solution (twice 5 min under gentle stirring). Saturate nonspecific sites again in PBT solution during 10 min. Load each slide with 60 μL of secondary antibody (produced in donkey) labeled with the appropriate fluorochrome and incubate again 45 min to 1 h at 37 °C in a dark humid container (do not let the slides dry). Wash with PBS-Tween 20 solution (twice 5 min under gentle stirring).

6. DNA staining step: (*Perform the following step in dim light*) Remove excess liquid by gently dabbing with a tissue (without completely drying the slides) and add 20 μL or a drop of mounting solution (Vectashield with a DNA counterstaining reagent). Cover with a glass cover slip, avoid bubbles as much as possible and remove the excess of mounting solution. Seal with nail varnish and let dry. The sealed slides can be stored at 4 °C for several days protected from light.

3.8.3 Metaphase Observation and Analysis

Observe each slide with an epifluorescent microscope under the 10× objective (to locate the area of the slide where the FISH signal is present and bright) and then with a 63× or 100× objective to capture the hybridization signal on the metaphase with the dedicated soft/hardware (*see* **Note 6**).

Figure 3a–d show typical examples of a whole-chromosome painting probe signal on metaphase spreads. FISH enables to confirm or identify which mouse chromosomes are involved in the rearrangement even when chromosome spreading is not top quality. We recommend observation of at least 30 metaphases to assess correctly the proportion of abnormal cells in a given cell culture.

4   Notes

1. Cell culture conditions for Embryonic Stem Cells (ESCs) and Epiblast Stem Cells (EpiSCs) are described elsewhere [7, 8]. Double synchronization with thymidine and synchronization with FdU is used on primary cell cultures only. No synchronization is needed for ESCs and EpiSCs because their cell cycle is very short but the proportion of dividing cells is enriched by a colchicine treatment during one hour thirty.

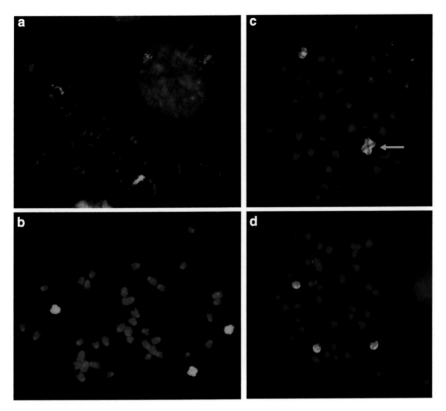

Fig. 3 Metaphases from two different mouse epiblast stem cell cultures with FISH signals (*green* fluorescence) of whole-chromosome painting probes coupled with FITC (specific of mouse chromosome 11). The DNA is counterstained in *red* with PI (**a**, **c** and **d**) or in *blue* (**b**) with DAPI. (**a**) Shows a normal metaphase with $2n = 40$ and two chromosomes 11. (**b–d**) Show two different kinds of chromosome 11 trisomy: in (**b**), cell line n°8.73 and (**d**), cell line n°3.83, the three chromosomes are independent and in (**c**), cell line n°3.83, two chromosomes 11 are fused (*arrow*) by their centromere (Robertsonian fusion)

2. The hypotonic solution used prior to metaphase spreading needs to be adapted for each cell culture type. Changes in the KCl versus serum proportion or incubation with KCl alone influence the quality of chromosome spreading and the success of the G-banding technique [9]. In our experience (*see* Table 1), the best KCl/serum proportion is 50/50 during 10 min for ESCs and EpiSCs and 80/20 for fibroblast cells. Thus, we recommend the test of various hypotonic treatments to determine optimal harvesting conditions for each cell culture.

3. For G-banding, the optimal trypsin treatment has to be found empirically by testing the same batch of slides on the same day with different incubation times. Room temperature is critical and temperature of the trypsin in the coplin jar must be monitored continuously to keep it at 11 °C. The method described

here is adapted to mouse chromosomes; other protocols have been described in ref. 9 for human cancer cells, in ref. 10 for human Embryonic Stem Cells (hESCs) and in refs. 6, 11 for porcine or bovine lymphocytes and it may be useful to test these for other cell types or species.

4. Chromosome painting probes are not commercially available for all species. Human whole-chromosomes probes (WCP) are now commercialized by several companies but good-quality WCP for mouse and rat can be purchased only from three suppliers: Metasystems (http://www.metasystems-international.com/xcyting-dna-probes), Chrombios (http://www.chrombios.com/cms/website.php?id=/en/index/fishproducts.htm&sid=hpvcuruuphr67ehkobhu10doe4) and Spectral imaging (http://www.spectral-imaging.com/reagents/single-probes). For species other than human, mouse, and rat, since the laboratory of M Ferguson-Smith (Cambridge Resource Centre for Comparative Genomics) is no longer operational, no commercial WCP are available and the only option is to prepare WCP by chromosome microdissection [12, 13]. On the whole, few commercial FISH probes for species other than human and mouse are available (*see* a review in ref. 14).

5. Primary and secondary antibodies can be purchased from various suppliers. We recommend rabbit anti-biotin antibodies (Bethyl) which in our hands gives no or little background (unlike streptavidin conjugated with a fluorophore) and *Jackson ImmunoResearch Inc.* for the secondary antibodies.

6. Nowadays, metaphases are captured with a digital camera linked to a software and/or hardware dedicated to cytogenetic analysis (classical and molecular). These image analysis systems are provided by several suppliers, i.e., Metasystem with Ikaros and Isis and Leica Biosystems with the CytoVision® platform. These hardware systems enable rapid reconstruction of karyotypes with user-friendly interfaces and tools.

7. Alternatively to thymidine, FdU can be used for cell synchronization. Divide the cell cultures as described in Subheading 3.1 into three dishes or flasks (H0). After 6 h (H6), add the FdU working solution at 10 μL/mL of culture medium. 14 h later (H20), incorporate the BrdU solution at 20 μL/mL of culture medium and harvest 6 h later (H26).

8. The RNase treatment is used to remove endogenous RNA and the pepsin treatment is used to decrease background noise and remove cytoplasmic/nucleoplasmic proteins that could prevent DNA-DNA hybridization and detection by the antibodies.

Acknowledgements

The authors wish to acknowledge Josette Catalan and the CeMEB Cytogenomics platform for their assistance in mouse stem cell karyotype analysis, Vincent Brochard and Sylvie Ruffini for their technical help in stem cells culturing and bovine fibroblast cultures respectively and Anne Calgaro and Alain Pinton for their advice on bovine karyotype analysis.

References

1. Fulka J Jr, Fulka H (2007) Somatic cell nuclear transfer (SCNT) in mammals: the cytoplast and its reprogramming activities. Adv Exp Med Biol 591:93–102

2. Guo J, Jauch A, Heidi HG et al (2005) Multicolor karyotype analyses of mouse embryonic stem cells. In Vitro Cell Dev Biol Anim 41:278–283

3. Sugawara A, Goto K, Sotomaru Y et al (2006) Current status of chromosomal abnormalities in mouse embryonic stem cell lines used in Japan. Comp Med 56:31–34

4. Sumner AT (1972) A simple technique for demonstrating centromeric heterochromatin. Exp Cell Res 75:304–306

5. Seabright M (1971) A rapid banding technique for human chromosomes. Lancet 2: 971–972

6. Ducos A, Pinton A, Yerle M et al (2002) Cytogenetic and molecular characterization of eight new reciprocal translocations in the pig species. Estimation of their incidence in French populations. Genet Sel Evol 34: 389–406

7. Jouneau A, Ciaudo C, Sismeiro O et al (2012) Naive and primed murine pluripotent stem cells have distinct miRNA expression profiles. RNA 18:253–264

8. Maruotti J, Dai XP, Brochard V et al (2010) Nuclear transfer-derived epiblast stem cells are transcriptionally and epigenetically distinguishable from their fertilized-derived counterparts. Stem Cells 28:743–752

9. MacLeod RA, Drexler HG (2013) Classical and molecular cytogenetic analysis. Methods Mol Biol 946:39–60

10. Meisner LF, Johnson JA (2008) Protocols for cytogenetic studies of human embryonic stem cells. Methods 45:133–141

11. Ducos A, Revay T, Kovacs A et al (2008) Cytogenetic screening of livestock populations in Europe: an overview. Cytogenet Genome Res 120:26–41

12. Kubickova S, Cernohorska H, Musilova P et al (2002) The use of laser microdissection for the preparation of chromosome-specific painting probes in farm animals. Chromosome Res 10:571–577

13. Pinton A, Ducos A, Yerle M (2003) Chromosomal rearrangements in cattle and pigs revealed by chromosome microdissection and chromosome painting. Genet Sel Evol 35:685–696

14. Rubes J, Pinton A, Bonnet-Garnier A et al (2009) Fluorescence in situ hybridization applied to domestic animal cytogenetics. Cytogenet Genome Res 126:34–48

Chapter 8

Treatment of Donor Cell/Embryo with Different Approaches to Improve Development After Nuclear Transfer

Eiji Mizutani, Sayaka Wakayama, and Teruhiko Wakayama

Abstract

The successful production of cloned animals by somatic cell nuclear transfer (SCNT) is a promising technology with many potential applications in basic research, medicine, and agriculture. However, the low efficiency and the difficulty of cloning are major obstacles to the widespread use of this technology. Since the first mammal cloned from an adult donor cell was born, many attempts have been made to improve animal cloning techniques, and some approaches have successfully improved its efficiency. Nuclear transfer itself is still difficult because it requires an accomplished operator with a practiced technique. Thus, it is very important to find simple and reproducible methods for improving the success rate of SCNT. In this chapter, we will review our recent protocols, which seem to be the simplest and most reliable method to date to improve development of SCNT embryos.

Key words Somatic cell nuclear transfer, Reprogramming, HDACi, Latrunculin A, SCNT embryo

1 Introduction

Animal cloning experiments were first performed in frogs [1, 2]. In 1997, Wilmut and Campbell succeeded in generating the first cloned sheep by transplanting nuclei from adult mammary cells [3, 4]. Soon after this breakthrough, cloning was achieved by somatic cell nuclear transfer (SCNT) in mice [5] and cattle [6]. Through careful technical refinements, it has now been achieved in many mammalian species [7–9]. However, the technique of SCNT is very difficult and the success rates of animal cloning are consistently low, especially in the mouse; typically, fewer than 3 % of cloned embryos develop to term [8, 10]. Many researchers have tried different methods to improve the success rate of animal cloning; for example, optimization of suitable recipient oocytes or donor cells [11, 12], using different stages of the cell cycle [13–19], using different strains of mice [20–22], or using cells with differing

Nathalie Beaujean et al. (eds.), *Nuclear Reprogramming: Methods and Protocols*, Methods in Molecular Biology, vol. 1222, DOI 10.1007/978-1-4939-1594-1_8, © Springer Science+Business Media New York 2015

differentiation status [23–25]. Optimizing oocyte activation methods [26, 27], changing the timing of removal of the metaphase II chromosomes from recipient oocytes [28], or increasing the cell numbers in SCNT embryos by aggregation [29] have also been reported. However, the success rate of animal cloning remains low.

The low efficiency of cloning technology is thought to be caused by incomplete reprogramming of the donor cell DNA. Thus, pretreatment of donor cells was expected to lead to an improvement in the success rate of animal cloning. Unfortunately, little improvement in the development of SCNT embryos has been achieved by treatment with drugs such as trichostatin A (TSA), a histone deacetylase inhibitor (HDACi), and/or 5-aza-2'-deoxycytidine, a DNA methyl-transferase inhibitor, to donor cells before NT [30, 31]. However, the possibility remains that appropriate treatment of donor cells may enhance cloning efficiency. In fact, Bui et al. reported that treatment of donor cells with germinal vesicle stage (GV) oocytes resulted in a threefold increase in the success rate of mouse cloning [32].

In contrast, treatment of oocytes with drugs after donor cell injection can markedly increase the cloning success rate. In 2006, there was a breakthrough in the SCNT technique. Kishigami et al. and Rybouchkin et al. found that TSA treatment of SCNT embryos could increase the cloning efficiency in mice sixfold [33, 34]. We then found that treatment with other HDACi, including scriptaid (SCR), suberoylanilide hydroxamic acid (SAHA), and oxamflatin, also enhanced the rate of development of SCNT embryos [35, 36]. These HDACi have been proposed to enhance the histone acetylation of SCNT embryos and the accessibility of the putative reprogramming factors by loosening the chromatin structure (Fig. 1a). HDACi treatment gives consistent results from mouse cloning (Fig. 1b). In fact, as shown in Fig. 1c, we have now succeeded in carrying out repeated recloning of mice over 25 generations [37]. In 2010, another promising approach for improving SCNT was reported. Inoue et al. found that X-linked genes were downregulated in SCNT embryos by ectopic expression of *Xist*, a noncoding RNA gene. They succeeded in increasing the birth rate of cloned mice tenfold by deletion of *Xist* from the active X chromosome in the donor genome [38], and they also succeeded in improving cloning efficiency by repression of ectopically expressed *Xist* with specific short interfering (si) RNA injection into SCNT embryos [39]. Although siRNA injection is a method that is applicable to wild-type mice, this technique did not enhance the efficiency of SCNT from female somatic cells [40]. These results indicated that correction of the aberrant epigenetic status of SCNT embryos is very important for the production of cloned offspring.

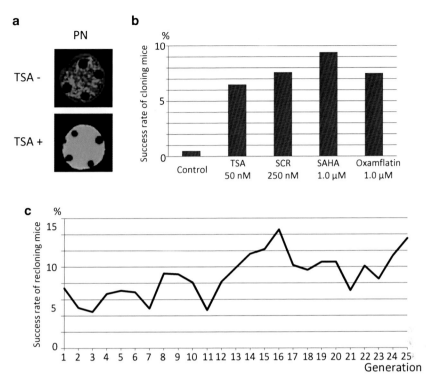

Fig. 1 Effects of TSA on the development of mouse SCNT embryos. (**a**) The level of histone H3 lysine 9 acetylation (aH3-K9) in the pronuclei of one-cell cloned embryos 10 h after activation. *Green*: aH3K9, *red*: nuclear membrane. (**b**) The rates of full-term development of SCNT embryos treated with HDACi TSA, SCR, SAHA, and oxamflatin in our laboratory. The concentrations of each HDACi are also shown. Treatment with all HDACi was for 10 h from the time of activation. (**c**) The success rate of recloning mice with TSA treatment

Most recently, we succeeded in increasing the rate of development of SCNT embryos to full term using a different approach. Terashita et al. rethought the conventional method for preventing pseudo-second polar body (PSPB) extrusion of reconstructed oocytes and used latrunculin A (LatA), an inhibitor of actin polymerization, instead of cytochalasin B (CB) or nocodazole [41]. Unlike CB or nocodazole, LatA did not impair F-actin polymerization or microtubule assembly, and treatment of SCNT embryos with 5 μM LatA resulted in twice the rate of full-term development obtained by treatment with CB (Fig. 2a, b). However, the SCNT procedure itself also needs more evaluation and refinement. If we can identify critical negative factors in the current protocols and correct them, we can increase the rate of successful full-term development of clones.

In this chapter, we will review our newly developed simple protocols to improve full-term development of SCNT embryos from cumulus cells using HDACi and/or LatA.

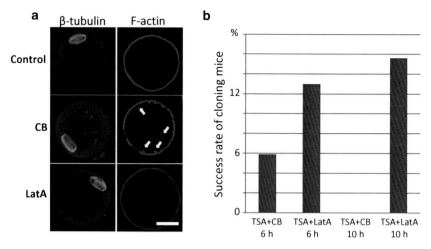

Fig. 2 Positive effects of LatA on mouse SCNT embryos. (**a**) Effects of CB or LatA treatment on the cytoskeleton. The *left images* show the localization of β-tubulin in a control MII stage oocyte and an MII stage oocyte treated with CB or LatA for 10 min. The *right images* show F-actin localization in IVF-generated control and partheno-genetic embryos treated with CB or LatA for 6 h. CB-treated oocytes had cytoplasmic asters and wider spindles compared with control and LatA-treated oocytes, and CB-treated embryos showed dot-like structures of F-actin (*arrows*). In contrast, the localization of β-tubulin and F-actin in LatA-treated oocytes or embryos was almost identical to those of controls. Bar = 30 μm. (**b**) The rate of development to full term of SCNT embryos treated with CB or LatA. LatA-treated SCNT embryos showed a higher rate of development than CB-treated embryos. Although no cloned mice were obtained from embryos treated with CB for 10 h, 10 h of continuous treatment with LatA enhanced the success rate of mouse cloning

2   Materials

2.1   Mice

F1 hybrid mice (2–3 months old) from strains such as B6D2F1 (C57BL/6 × DBA/2) or B6C3F1 (C57BL/6 × C3H/He) are used for recipient metaphase II oocyte collection (*see* **Note 1**). Hybrid strains such as B6D2F1, B6C3F1, or B6 × 129F1 are usually used as donors because they provide much better donor cells for the production of cloned mice. The ICR (CD-1) strain of mice is used for pseudopregnant surrogate mothers or foster mothers and for providing vasectomized males.

2.2   Small Lab Materials

1. Disposable consumables: 1 ml syringes with needles for hormone injections, 6 cm culture dishes, permanent marker, embryo transfer pipettes.

2. Equipment: dissecting microscope, scissors, tweezers.

2.3   Chemicals and Sources

1. Hormones: 50 IU/mL equine chorionic gonadotropin (eCG), 50 IU/mL human chorionic gonadotropin (hCG). Both eCG and hCG are dissolved in saline and stored at −40 °C until use.

2. 500 μg/mL Cytochalasin B (CB) (100× stock solution): CB is dissolved in dimethyl sulfoxide (DMSO), aliquoted into small tubes (10–20 μL), and stored at −40 °C.

3. 100 mM $SrCl_2$ (20× stock solution): $SrCl_2$ is dissolved in ultrapure water and stored in aliquots at room temperature.

4. 200 mM Ethylene glycol tetraacetic acid (EGTA) (100× stock solution): EGTA is dissolved in ultrapure water and stored in aliquots at 4 °C.

5. 10 % Hyaluronidase (100× stock solution): Hyaluronidase is dissolved in M2 medium and stored at –40 °C.

6. HDACi 200× stock solutions: 10 μM TSA, 25 μM SCR, 100 μM SAHA, 100 μM oxamflatin. A 200× stock solution of each HDACi is prepared by dissolving it in DMSO at a suitable concentration and storing at –80 °C until use.

7. 1 mM LatA (200× stock solution): LatA is dissolved in DMSO and stored at –40 °C until use.

8. Other reagents: mineral oil and mercury.

2.4 Media

1. KSOM: 0.5 mg/mL PVA, 95 mM NaCl, 2.5 mM KCl, 0.3 mM KH_2PO_4, 0.2 mM $MgSO_4$, 0.2 mM Na pyruvate, 25 mM $NaHCO_3$, 10 mM Na lactate, 0.01 mM EDTA, 2.8 mM glucose, 1.0 mM glutamine, 1.7 mM $CaCl_2$, 0.5× nonessential amino acids, 0.5× essential amino acids, 1 mg/mL bovine serum albumin. All components are dissolved in ultrapure water.

2. M2 medium: 95 mM NaCl, 4.8 mM KCl, 1.2 mM KH_2PO_4, 1.2 mM $MgSO_4$, 23 mM Na lactate, 0.3 mM Na pyruvate, 5.6 mM glucose, 4.15 mM $NaHCO_3$, 1.7 mM $CaCl_2$, 20.9 mM HEPES. All reagents are dissolved in ultrapure water and the pH adjusted to 7.3–7.4.

3. 0.1 % hyaluronidase M2 medium: 2 μL of hyaluronidase 100× stock solution, 198 μL of M2 medium.

4. Polyvinylpyrrolidone (PVP) medium. PVP (360 kDa) is dissolved in M2 medium at 10–12 %, aliquoted into small tubes (50–100 μL), and stored at –40 °C.

5. M2 + CB medium: 2 μL of CB 100× stock solution, 198 μL of M2 medium.

6. Activation medium for HDACi treatment: 10 μL of $SrCl_2$ 20× stock solution, 2 μL of EGTA 100× stock solution, 2 μL of CB 100× stock solution, 1 μL of HDACi 200× stock solution, 185 μL of KSOM.

7. HDACi medium: 1 μL of HDACi 200× stock solution, 199 μL of KSOM.

8. Activation medium for LatA and HDACi treatment: 10 μL of $SrCl_2$ 20× stock solution, 2 μL of EGTA 100× stock solution, 1 μL of LatA 200× stock solution, 1 μL of HDACi 200× stock solution, 186 μL of KSOM.

2.5 Tools for Nuclear Transfer

1. Inverted microscope with Hoffman modulation contrast optics (*see* **Note 2**).

2. Micromanipulator set.

3. Microforge.

4. Pipette puller and glass micromanipulation pipettes (*see* **Note 3**).

5. Piezo impact drive system.

6. Warming plate or warming stage on the microscope set at 37 °C.

7. Humidified incubator: 37 °C, 5 % CO_2.

3   Methods

3.1 Preparation of Mice for Collection of Recipient Oocytes and Donor Cumulus Cells

Female mice are induced to superovulate by administration of hormones; 5 IU eCG is injected into the abdominal cavity 3 days before an experiment, with 5 IU hCG 48 h later (1 day before an experiment). Usually we inject mice at 5–6 pm.

3.2 Preparation of Dishes for Handling Oocytes or Embryos

1. Culture and oocyte-washing dishes: before starting any experiment, the dishes for oocyte collection must be prepared. Place a number of drops of KSOM (about 15 μL) on a 6 cm culture dish, cover with mineral oil, and place in the incubator. The same type of dishes can be used for oocyte washing during manipulation procedures and for long-term cultures of cloned embryos.

2. Dishes for the micromanipulation chamber: Place droplets (~10 μL) of the three media (M2 for injection, M2 + CB for enucleation, PVP for donor cell suspension and washing needles) on the lid of a 10 cm dish and cover with mineral oil. To distinguish each medium, draw lines on the back of the dish.

3. Dishes for oocyte activation: Place droplets (~15 μL) of activation media on a 6 cm dish and cover with mineral oil.

3.3 Collection of Recipient Oocytes and Donor Cumulus Cells

1. Collect oocyte–cumulus cell complexes from the oviduct ampullae 14–15 h after hCG injection (usually, we collect oocytes at 8–9 am) and move them into a droplet of 0.1 % hyaluronidase M2 medium. After 5 min, pick up good-quality oocytes, wash them three times in M2 medium, place in the drops of KSOM, and incubate until use.

2. The remaining cumulus cells are used as donor cells. Cumulus cells are transferred to PVP droplets for donor cell suspension (*see* **step 2** in Subheading 3.2) and mixed with a glass pipette or tweezers.

3.4   Preparation of Micropipettes

1. All micropipettes for NT can be made in the laboratory using a pipette puller and microforge. For the holding pipette, the outside diameter (OD) should be equal to or a little smaller than that of an oocyte (e.g., OD 70 μm, inner diameter (ID) 10 μm). The ID of the enucleation pipette should be 7–9 μm. The ID of the injection pipette depends on donor cell type. The best ID for cumulus cells is 5–6 μm.

2. The tip section of the pipettes (about 300 μm back from the end) is bent at 15–20°. A small amount of mercury is back-loaded into the enucleation and injection pipettes before use (*see* **Note 4**).

3.5   Nuclear Transfer (Also See Chapter 3)

1. Place one group of oocytes (between 10 and 25) in a droplet of M2 + CB medium on the micromanipulation chamber and wait at least 5 min.

2. Remove the MII spindle (a clear zone in the oocyte observed using Nomarski or Hoffman optics) with a minimal volume of cytoplasm, then wash the enucleated oocytes at least three times in KSOM to remove the CB completely and return them to the incubator in droplets of KSOM for at least 30 min until donor cell injection (*see* **Note 5**).

3. Add a small volume of donor cell suspension into the PVP medium and mix completely. Place one group of enucleated oocytes into M2 medium.

4. Break the membrane of each donor cell by gently aspirating it in and out of the injection pipette and inject donor nuclei into the cytoplasm of the enucleated oocytes (*see* **Note 6**).

5. Injected oocytes should be kept in the injection droplet for at least 10 min before returning them to KSOM in the incubator (*see* **Note 7**).

3.6   Treatment with HDACi to Improve Development of SCNT Embryos (See Note 8, Fig. 3)

1. Transfer the reconstructed oocytes to droplets of activation medium for HDACi treatment, wash twice, then culture for 6 h in the incubator at 37 °C under 5 % CO_2.

2. After 6 h culture in the activation medium for HDACi treatment, embryos should be transferred to droplets of HDACi medium and washed twice before being cultured. If NT and oocyte activation are performed properly, a few pseudopronuclei should be observed in reconstructed embryos at this time (*see* **Note 9**). To enhance genomic reprogramming, culture in HDACi medium for up to 3–4 h.

3. Wash the cloned embryos in KSOM and transfer to new droplets of the same medium prepared in another dish, then continue culturing (*see* **Note 10**).

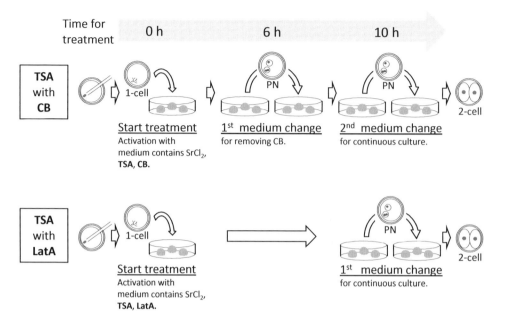

Fig. 3 Time frame for treatment with TSA combined with CB or LatA. The toxicity of LatA was much lower than that of CB and there was no need to wash the embryos at the 6 h time point after activation. This new simplified and effective activation method to improve development of SCNT embryos will assist in the widespread use of SCNT technology

3.7 Treatment with LatA and HDACi to Improve Development of SCNT Embryos (See Note 8, Fig. 3)

1. Transfer the reconstructed oocytes to droplets of activation medium for LatA and HDACi treatment and wash twice, then culture for 10 h in the incubator at 37 °C in 5 % CO_2.

2. Wash the cloned embryos in KSOM and transfer to new droplets of the same medium prepared in another dish, then continue culturing (*see* **Note 11**).

3.8 Embryo Transfer

1. Two-cell stage cloned embryos (24 h after activation) or 4- to 8-cell embryos (48 h after activation) can be transferred to oviducts of pseudopregnant female mice at 0.5 days postcopulation (dpc). Morulae/blastocysts (72 h after activation) should be transferred into the uterus of 2.5 dpc pseudopregnant female mice. Thus, mating of ICR females with vasectomized male mice must be timed correctly, depending on the purpose of your experiment.

2. At 18.5 or 19.5 dpc, euthanize the surrogate mothers and subject them to Cesarean section. Remove the uterus from the abdomen and dissect out the cloned pups with their placentas. Wipe away the amniotic fluid from the skin, mouth, and nostrils and stimulate the pups to breathe by rubbing their back or pinching them gently with blunt forceps, and warm to 37 °C (*see* **Note 12**).

3. After recovering, the cloned pups are transferred to the cage of a foster mother with natural delivery. Take some soiled bedding from the cage and nestle the cloned pups in the bedding material so that they take on the odor of the bedding. Remove some pups from the foster mother's litter and then mix the cloned pups with the foster female's pups.

4   Notes

1. The mouse strain used for recipient oocytes is indeed important for successful NT experiments. In B6D2F1 oocytes, it is easy to find the MII spindle and they are robust enough to withstand in vitro manipulation and culture. It is possible to use inbred stains, such as C57BL/6 or C3H/He, but with a lower success rate for producing cloned mice than with hybrid strains.

2. An inverted microscope can also be used with Nomarski differential interference optics. In this case, a glass-bottomed dish must be used to obtain the best resolution.

3. Micropipettes for micromanipulation can be ordered from several companies (e.g., Prime Tech Ltd. http://www.primetech-jp.com/en/01products/psk.html).

4. Mercury is volatile and is toxic if absorbed by breathing or through the skin. Wear appropriate gloves and always handle mercury in a working fume hood. Use appropriate safety handling conditions, as recommended by your Institutional Safety Officer.

5. Contamination with CB affects the survival rate of oocytes after injection. If the CB is not washed out completely, many oocytes will lyse during microinjection.

6. Do not apply the piezo pulse until the tip of the injection pipette reaches the opposite wall of the oocyte. This is important to enable the oocytes to survive after injection.

7. This step is also important for the survival of injected oocytes. The oocyte membrane must be allowed to recover before returning it to the incubator.

8. Subheadings 3.6 and 3.7 are the simplest and most reliable methods to date to enhance mouse cloning efficiency. Inhibition of PSPB extrusion is important when using G0/G1 phase somatic cells as donor nuclei. If G2/M phase cells are used as donor cells, CB or LatA must be omitted from the activation medium to allow the oocyte to eject the PSPB so that the embryos can remain diploid; in this case, you should choose Subheading 3.6 and use activation medium for HDACi treatment without CB.

9. The most probable reason for failure to form pseudopronuclei is a failure to break the donor cell membranes when they are injected into oocytes. To avoid this problem, the inner diameter of the injection pipette must be smaller than the donor cell. Apply a few piezo pulses to break the donor cell membrane when you pick it up.

10. To avoid carry-over of the chemicals contained in activation media into the medium for culture, all embryos should be washed at least three times and moved to different culture dishes for long-term culture.

11. Interestingly, unlike treatment with CB, treatment with LatA for 10 h gives a higher success rate for mouse cloning than did 6 h treatment. This procedure allows us to skip the stage of changing the medium at 6 h and to simplify the mouse SCNT procedure.

12. If you have no success in generating cloned mice, check the media used in the experiments. However, the most important thing is your technical skill. Do not give up!

References

1. Briggs R, King TJ (1952) Transplantation of living nuclei from blastula cells into enucleated frogs' eggs. (Translated from eng). Proc Natl Acad Sci U S A 38(5):455–463

2. Gurdon JB, Uehlinger V (1966) "Fertile" intestine nuclei. (Translated from eng). Nature 210(5042):1240–1241

3. Campbell KH, McWhir J, Ritchie WA, Wilmut I (1996) Sheep cloned by nuclear transfer from a cultured cell line. (Translated from eng). Nature 380(6569):64–66

4. Wilmut I, Schnieke AE, McWhir J, Kind AJ, Campbell KH (1997) Viable offspring derived from fetal and adult mammalian cells. (Translated from eng). Nature 385(6619):810–813

5. Wakayama T, Perry AC, Zuccotti M, Johnson KR, Yanagimachi R (1998) Full-term development of mice from enucleated oocytes injected with cumulus cell nuclei. (Translated from eng). Nature 394(6691):369–374

6. Kato Y et al (1998) Eight calves cloned from somatic cells of a single adult. (Translated from eng). Science (New York, NY) 282(5396): 2095–2098

7. Cibelli J (2007) Developmental biology. A decade of cloning mystique. (Translated from eng). Science 316(5827):990–992

8. Thuan NV, Kishigami S, Wakayama T (2010) How to improve the success rate of mouse cloning technology. (Translated from eng). J Reprod Dev 56(1):20–30

9. Momosaki S et al (2010) PK11195 might selectively suppress the quinolinic acid-induced enhancement of anaerobic glycolysis in glial cells. Brain Res 1340:18–23

10. Wakayama T (2007) Production of cloned mice and ES cells from adult somatic cells by nuclear transfer: how to improve cloning efficiency? (Translated from eng). J Reprod Dev 53(1):13–26

11. Wakayama T, Yanagimachi R (2001) Mouse cloning with nucleus donor cells of different age and type. (Translated from eng). Mol Reprod Dev 58(4):376–383

12. Hiiragi T, Solter D (2005) Reprogramming is essential in nuclear transfer. (Translated from eng). Mol Reprod Dev 70(4):417–421

13. Amano T, Tani T, Kato Y, Tsunoda Y (2001) Mouse cloned from embryonic stem (ES) cells synchronized in metaphase with nocodazole. (Translated from eng). J Exp Zool 289(2): 139–145

14. Cibelli JB et al (1998) Cloned transgenic calves produced from nonquiescent fetal fibroblasts. (Translated from eng). Science (New York, NY) 280(5367):1256–1258

15. Kasinathan P, Knott JG, Wang Z, Jerry DJ, Robl JM (2001) Production of calves from G1 fibroblasts. (Translated from eng). Nat Biotechnol 19(12):1176–1178

16. Ono Y, Shimozawa N, Ito M, Kono T (2001) Cloned mice from fetal fibroblast cells arrested at

metaphase by a serial nuclear transfer. (Translated from eng). Biol Reprod 64(1):44–50

17. Ono Y et al (2001) Production of cloned mice from embryonic stem cells arrested at metaphase. (Translated from eng). Reproduction (Cambridge, England) 122(5):731–736

18. Tani T, Kato Y, Tsunoda Y (2003) Reprogramming of bovine somatic cell nuclei is not directly regulated by maturation promoting factor or mitogen-activated protein kinase activity. (Translated from eng). Biol Reprod 69(6):1890–1894

19. Wakayama T, Rodriguez I, Perry AC, Yanagimachi R, Mombaerts P (1999) Mice cloned from embryonic stem cells. (Translated from eng). Proc Natl Acad Sci U S A 96(26):14984–14989

20. Inoue K et al (2003) Effects of donor cell type and genotype on the efficiency of mouse somatic cell cloning. (Translated from eng). Biol Reprod 69(4):1394–1400

21. Ogura A et al (2000) Production of male cloned mice from fresh, cultured, and cryopreserved immature Sertoli cells. (Translated from eng). Biol Reprod 62(6):1579–1584

22. Saito M, Saga A, Matsuoka H (2004) Production of a cloned mouse by nuclear transfer from a fetal fibroblast cell of a mouse closed colony strain. (Translated from eng). Exp Anim 53(5):467–469

23. Kato Y et al (2004) Nuclear transfer of adult bone marrow mesenchymal stem cells: developmental totipotency of tissue-specific stem cells from an adult mammal. (Translated from eng). Biol Reprod 70(2):415–418

24. Mizutani E et al (2006) Developmental ability of cloned embryos from neural stem cells. (Translated from eng). Reproduction (Cambridge, England) 132(6):849–857

25. Inoue K et al (2006) Inefficient reprogramming of the hematopoietic stem cell genome following nuclear transfer. (Translated from eng). J Cell Sci 119(Pt 10):1985–1991

26. Kishikawa H, Wakayama T, Yanagimachi R (1999) Comparison of oocyte-activating agents for mouse cloning. (Translated from eng). Cloning 1(3):153–159

27. Wakayama T, Yanagimachi R (2001) Effect of cytokinesis inhibitors, DMSO and the timing of oocyte activation on mouse cloning using cumulus cell nuclei. (Translated from eng). Reproduction (Cambridge, England) 122(1):49–60

28. Wakayama S, Cibelli JB, Wakayama T (2003) Effect of timing of the removal of oocyte chromosomes before or after injection of somatic nucleus on development of NT embryos. (Translated from eng). Cloning Stem Cells 5(3):181–189

29. Boiani M, Eckardt S, Leu NA, Scholer HR, McLaughlin KJ (2003) Pluripotency deficit in clones overcome by clone-clone aggregation: epigenetic complementation? (Translated from eng). EMBO J 22(19):5304–5312

30. Enright BP, Kubota C, Yang X, Tian XC (2003) Epigenetic characteristics and development of embryos cloned from donor cells treated by trichostatin A or 5-aza-2'-deoxycytidine. Biol Reprod 69(3):896–901

31. Shi W et al (2003) Induction of a senescent-like phenotype does not confer the ability of bovine immortal cells to support the development of nuclear transfer embryos. Biol Reprod 69(1):301–309

32. Bui HT et al (2008) The cytoplasm of mouse germinal vesicle stage oocytes can enhance somatic cell nuclear reprogramming. Development 135(23):3935–3945

33. Kishigami S et al (2006) Significant improvement of mouse cloning technique by treatment with trichostatin A after somatic nuclear transfer. (Translated from eng). Biochem Biophys Res Commun 340(1):183–189

34. Rybouchkin A, Kato Y, Tsunoda Y (2006) Role of histone acetylation in reprogramming of somatic nuclei following nuclear transfer. (Translated from eng). Biol Reprod 74(6):1083–1089

35. Van Thuan N et al (2009) The histone deacetylase inhibitor scriptaid enhances nascent mRNA production and rescues full-term development in cloned inbred mice. (Translated from eng). Reproduction (Cambridge, England) 138(2):309–317

36. Ono T et al (2010) Inhibition of class IIb histone deacetylase significantly improves cloning efficiency in mice. Biol Reprod 83(6):929–937

37. Wakayama S et al (2013) Successful serial recloning in the mouse over multiple generations. Cell Stem Cell 12(3):293–297

38. Inoue K et al (2010) Impeding Xist expression from the active X chromosome improves mouse somatic cell nuclear transfer. Science (New York, NY) 330(6003):496–499

39. Matoba S et al (2011) RNAi-mediated knockdown of Xist can rescue the impaired postimplantation development of cloned mouse embryos. Proc Natl Acad Sci U S A 108(51):20621–20626

40. Oikawa M et al (2013) RNAi-mediated knockdown of Xist does not rescue the impaired development of female cloned mouse embryos. J Reprod Dev 59(3):231–237

41. Terashita Y et al (2012) Latrunculin A can improve the birth rate of cloned mice and simplify the nuclear transfer protocol by gently inhibiting actin polymerization. Biol Reprod 86(6):180

Chapter 9

Fluorescent Immunodetection of Epigenetic Modifications on Preimplantation Mouse Embryos

Claire Boulesteix and Nathalie Beaujean

Abstract

A common problem in research laboratories that study the mammalian embryo after nuclear transfer is the limited supply of material. For this reason, new methods are continually developed, and existing methods for cells in culture are adapted to suit this peculiar experimental model. Among them is the fluorescent immunodetection.

Fluorescent immuno-detection on fixed embryos is an invaluable technique to detect and locate proteins, especially nuclear ones such as modified histones, in single embryos thanks to its specificity and its sensitivity. Moreover, with specific fixation procedures that preserve the 3D shape of the embryos, immunostaining can now be performed on whole-mount embryos. Target proteins are detected by specific binding of first antibody usually nonfluorescent, and revealed with a second antibody conjugated with a fluorochrome directed specifically against the host animal in which the first antibody was produced. The result can then be observed on a microscope equipped with fluorescent detection. Here, we describe the 3D fluorescent immunodetection of epigenetic modifications in mouse embryos. This procedure can be used on nuclear transferred embryos but also on in vivo-collected, in vitro-developed and in vitro-fertilized ones.

Key words Embryo, Mouse, Development, Epigenetic, Reprogramming, Immunodetection, Fluorescence, Microscopy

1 Introduction

During normal development the two very specialized gametes, the sperm and the oocyte, undergo profound changes. They lose their characteristic nuclear and chromatin configuration giving way to the establishment of an embryo-specific chromatin and nuclear organization. It is exactly this distinctive embryonic configuration that should be achieved by the donor nucleus after nuclear transfer (NT) into an enucleated oocyte. This conversion of a differentiated nuclear organization into an embryonic one is forced by the intimate contact between the donor cell chromatin and the oocyte cytoplasm, allowing its "reprogramming". Although the mechanisms of this reprogramming remain poorly understood, it appears that the particular chromatin landscape established after nuclear

Nathalie Beaujean et al. (eds.), *Nuclear Reprogramming: Methods and Protocols*, Methods in Molecular Biology, vol. 1222, DOI 10.1007/978-1-4939-1594-1_9, © Springer Science+Business Media New York 2015

transfer is crucial for embryonic genome expression and for further development [1–4].

Whereas the genetic information provides the framework for the manufacture of RNAs and proteins, chromatin structure and nuclear architecture indeed control the accessibility of proteins to the DNA, especially transcription factors and RNA polymerase, and thereby gene expression [5]. It involves several mechanisms such as epigenetic modifications (DNA methylation or histone post-translational modifications for example) that can be examined by immunostaining [6].

During the past decade, 3D immunodetection on fixed NT embryo has therefore provided tremendous insight into the sub-nuclear positioning of proteins during these very first stages of development, and into mechanisms—such as the epigenetic ones modifying the degree of chromatin compaction—involved in the reprogramming of the donor cell [7].

Fluorescent immunostaining allows the simultaneous detection and localization of two or three epigenetic modifications by combining primary antibodies produced in different species, and secondary antibodies with different fluorochromes. Moreover, immunodetection makes it possible to have results even with a single embryo at the 1-cell stage, thereby reducing the problem encountered by the limited availability of embryos after nuclear transfer.

The immunofluorescent-staining protocols usually used on thin cells in culture tend to flatten the specimens and are not optimal for the analysis of embryos. However by using alternative fixation procedures ensuring that the specimens maintain the 3D structure they had in vivo, 3D data sets can now be obtained on preimplantation embryos with a preserved shape [8, 9].

By combining this technique with high resolution microscopy—Structured Illumination Microscope or Confocal Laser Scanning Microscope- and precise image analysis it is also possible to quantify the epigenetic modification of interest and then to describe its evolution not only in term of presence or absence, but also in terms of localization and quantity during preimplantation development [4, 10–13].

Here, we describe the 3D fluorescent immunodetection of epigenetic modifications in mouse embryos with (1) the fixation of the embryos, (2) the immunostaining itself, and (3) the observation procedure.

2   Materials

For all the procedures some basic equipment is required (Fig. 1).

1. Dissecting microscope with appropriate working distance.

2. Heating block set at 27 °C (*see* **Note 1**).

3. Aspirator tube: about 40 cm length (*see* **Note 2**) (Fig. 2).

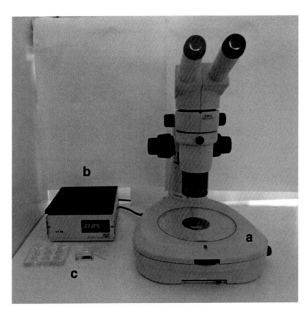

Fig. 1 Basic equipment required. Performing immunostaining on whole-mount preimplantation embryos requires (**a**) a dissecting microscope with appropriate working distance, (**b**) an heating block set at 27 °C and (**c**) embryo glass dishes (with one or several wells) with glass lids

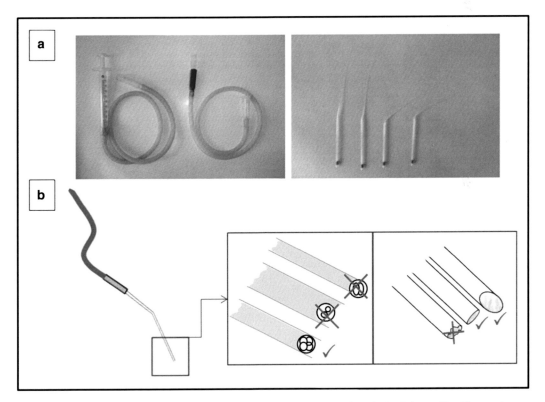

Fig. 2 Aspirator tube and glass manipulation pipettes. (**a**) Examples of aspirator tubes with either syringe or mouth control and of some glass manipulation pipettes (straight and oblique). (**b**) Choice of the right diameter for the manipulation of the embryos: it should fit exactly to the diameter of the embryos and have smooth extremity. If necessary the diameter can be increased by breaking carefully (with your nails or with a diamond pen) the extremity of the glass pipette

4. Thin glass manipulation pipettes for embryos, usually home made in the flame with glass capillaries or Pasteur pipettes (*see* **Note 3**) (Fig. 1).

5. Embryo glass dishes with glass lids (*see* **Note 4**) (Fig. 1).

2.1 Embryo Fixation

1. pH indicator strips.

2. Fixative: 20 % paraformaldehyde (PFA) solution. Upon opening of the stock solution, adjust pH to ~7.0 with 0.2 N NaOH if necessary and store maximum 1 month at 4 °C with light protection (aluminum foil for example). Before each experiment pH should be checked with a pH indicator strip: if lower than 7.0 or upper to 7.4, throw it away.

3. Phosphate Buffered Saline (PBS): prepare according to manufacturer's instructions, autoclave, and store at 4 °C. Filter through a 0.22-μm sterile cellulose acetate membrane before use.

4. Parafilm.

2.2 Immunostaining

Prepare all solutions using ultrapure water (18.2 MΩ cm resistivity at 25 °C).

1. Phosphate Buffered Saline (PBS): same preparation as above (**item 3** in Subheading 2.1).

2. Permeabilization stock solution: 10 % Triton X-100 stock solution is prepared with ultrapure water (w/v) and mixed slowly to avoid bubbles. Store at 4 °C for maximum 1 month.

3. PBS-BSA 2 % solution: dissolve embryo/culture tested BSA powder slowly in PBS (w/v) and filter through 0.22-μm sterile cellulose acetate membranes. Aliquots of 1 or 2 mL can be kept at –20 °C for several months and thawed just before use. After thawing filter through a 0.22-μm sterile cellulose acetate membrane, store at 4 °C.

4. Primary and secondary antibodies (*see* **Note 5**).

5. Aluminum foil.

6. Petri dishes (35 or 60 mm diameter).

7. 0.5 or 1.5 mL micro test tubes.

8. Mineral oil, embryo tested: keep at room temperature and protect from direct light.

9. 37 °C oven or incubator (*see* **Note 6**).

10. Nucleic acid stain (*see* **Note 7**).

11. Mounting anti-fading medium.

12. SuperFrost Plus slides.

13. Glass cover slips (*see* **Note 8**).

14. Clear nail polish.

15. Thin micro-spatula.

2.3 Microscope ***Observation***	High resolution microscopy—such as Structured Illumination Microscope or Confocal Laser Scanning Microscope—is highly recommended.

3 Methods

3.1 Embryo Fixation	1. Label the glass dishes and lids with appropriate signs in order to identify the different groups (control, NT etc…), date and size ($n = \dots$).
	
	2. Choose between the two types of fixation:
	
	   – *A short one*: 20 min at room temperature (22–25 or 27 °C on the heating plate) with 2 % PFA freshly diluted in Phosphate Buffered Saline PBS, if you plan to perform immunostaining immediately after.
	
	   – *A long one*: overnight at +4 °C with 4 % PFA, freshly diluted in Phosphate Buffered Saline PBS, if you plan to perform immunostaining the day after. If necessary keep the embryos in PFA for up to 1 week—in that case, change the PFA 4 % every 2 days (*see* **Note 9**).
	
	3. Prepare in each glass dish a volume of ~500 µL PFA, diluted with PBS.
	
	4. With a thin glass manipulation pipette transfer the embryos group by group in the corresponding glass dish (*see* **Note 10**) (Fig. 3), directly from the collection/culture medium to the fixative solution, with no rinse step (*see* **Note 11**).

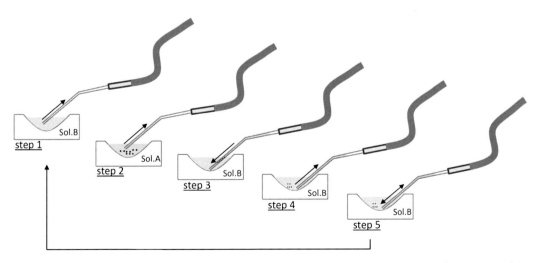

Fig. 3 Embryo transfer step by step. The quality of the immunostaining resides in the perfect contact of the embryos with each incubation solution. For each transfer from solution A to solution B, follow the steps as described: *Step 1*: aspirate solution B to halfway up the glass manipulation pipette. *Step 2*: aspirate five embryos, in a minimum volume of the solution A. *Step 3*: breathe out the embryos in solution B and discard the rest of solution A. *Step 4*: aspirate solution B to halfway up the glass manipulation pipette. *Step 5*: move the embryos around in the glass dish several time. Repeat the five steps as many times as necessary, taking five embryos each time

5. Cover the glass dishes with the glass lids; in the case of a long fixation wrap each dish with parafilm to avoid drying out of the fixative at 4 °C.

6. Incubate in PFA, either 20 min at room temperature (or 27 °C on the heating plate) or overnight at +4 °C.

3.2 Immunostaining

All steps are performed at room temperature (22–25 or 27 °C on the heating plate) with a total volume of 500 μL per glass dish, unless otherwise mentioned. Between each step diligently follow hazardous waste recycling regulations. Carefully wash each glass dish with distilled water and dry it with a nonfluffy rag.

1. Transfer the embryos with the glass embryo manipulation pipettes from the fixative into PBS:

   (a) if PFA incubation was ≤30 min, 2 times 5 min PBS rinse is enough

   (b) if PFA incubation was longer, wash for at least 30 min (3 times 10 min).
   In both cases, and at each further step, embryos have to be aspirated in the pipette with a minimum amount of solution, breathed out in the next solution and moved within the glass dish several times (*see* **Note 10**) (Fig. 3).

2. For permeabilization, transfer the embryos and incubate 30 min in 0.5 % TritonX100, freshly diluted in PBS (mix gently to avoid bubbles) (*see* **Note 12**).

3. Saturation of nonspecific sites is performed by 1 h incubation of the embryos in PBS containing 2 % BSA (PBS–BSA 2 %) (*see* **Note 13**).

4. Dilute the primary antibody with PBS containing 2 % BSA in a micro test tube (*see* **Note 14**). You will need a total of 20 μl per group + 20 μl for equilibration.

5. In a Petri dish prepare 20 μl drops of the primary antibody for each group and add one for equilibration (*see* **Note 15**). Cover the drops with mineral oil (Fig. 4).

6. Embryos are transferred group by group in the primary antibody: first in the equilibration drop and then in their respective antibody incubation drops.

7. Incubation in the first antibody can be done either 1 h at 37 °C or overnight at 4 °C (*see* **Note 16**). If you are using fluorescent primary antibodies, directly go to **step 10** after this incubation.

8. Transfer the embryos in glass dishes of PBS. Wash 3 times 10 min to remove antibody excess (*see* **Note 17**).

9. Second antibody incubation: dilute in PBS–BSA 2 % (*see* **Note 18**), prepare 20 μl drops under mineral oil in a Petri dish

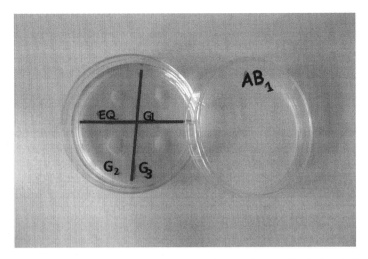

Fig. 4 Antibody incubation. For primary and secondary antibody incubations prepare for each group (G1, G2, G3 etc…) 20 µl drops of the antibody (diluted with PBS containing 2 % BSA) and add one drop for equilibration (EQ). Cover the drops with mineral oil and transfer the embryos group by group: first in the equilibration drop and then in their respective antibody incubation drops

(*see* **Note 15**) and transfer the embryos group by group as above (first in the equilibration drop and then in their respective antibody incubation drops), incubate 1 h at room temperature (22–25 or 27 °C on the heating plate).

From this step, to the end, always carefully protect dishes from light, with aluminum foil for example.

10. Transfer the embryos in glass dishes of PBS. Wash 3 times 10 min to remove antibody excess (*see* **Note 17**).

11. Embryos can be transferred in 2 % PFA (diluted in PBS) for 30 min + washed once with PBS (2 min) (*see* **Note 19**).

12. Transfer the embryos in glass dishes containing a nucleic acid stain diluted in PBS and incubate for 20 min at 37 °C (*see* **Note 20**).

3.3 Mounting the Embryos on Slides

1. Prepare SuperFrost Plus slides with identification of the experiment on the side (date/species/embryonic stage/number of embryos and antibody used) and as many areas as groups drawn on the back on the slide with a permanent pen (*see* **Note 21** and Fig. 5a).

2. Load a 100 µl pipetman with 40 µL of the antifading agent (*see* Fig. 5b).

3. Aspirate the DNA staining solution to halfway up the glass manipulation pipette and then aspirate five embryos of the first group, in a minimum volume of solution (*see* **Note 22** and Fig. 5d step 1).

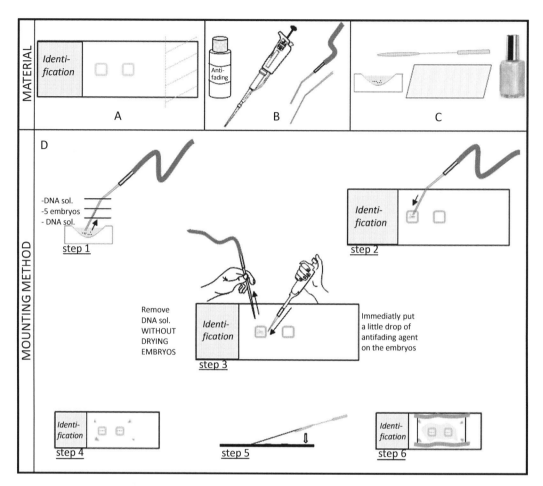

Fig. 5 Mounting of the immunostained embryos on slides. Mounting the embryos on slides has to be done carefully in order to preserve the 3D-shape of the samples. First prepare the following material: (**a**) SuperFrost Plus slides with identification of the experiment on the side (date/species/embryonic stage/number of embryos and antibody used) and as many areas as groups drawn on the back on the slide with a permanent pen (do not use the extremity of the slide: when using an inverted microscope for observations the objective could be damaged); (**b**) 100 μl pipetman ready to use with 40 μL of the antifading agent and two glass manipulation pipettes (a perfectly calibrated and a very thin one with a diameter smaller than the embryos); (**c**) the immunostained embryos in their glass dishes, a glass cover slip, a micro-spatula, and clear nail polish to secure the cover slip into place. Then follow the mounting steps: *Step 1*: aspirate the DNA staining solution to halfway up the glass manipulation pipette and then aspirate five embryos of the first group, in a minimum volume of the DNA staining solution. *Step 2*: breathe out all the embryos, trying to put them in line in the delimited area on the slide. Repeat *steps 1* and *2* as many times as necessary, taking five embryos from the first group each time. *Step 3*: in one hand, with a very thin pipette aspirate as much solution around the embryos as possible, however be particularly careful to not dry the embryos/in the other hand, be ready, with the pipetman containing the antifading agent to put down a little drop of antifading agent on embryos immediately after the removal of the DNA staining solution. Repeat all the steps for the other groups of embryos. *Step 4*: put a little drop of clear nail polish on each corner of the slide, next to the groups, in order to preserve the 3D-shape of the embryos. *Step 5*: carefully put the cover slip on the slide helping yourself with a micro-spatula, starting from one side and getting down the other side slowly. *Step 6*: secure with clear nail polish around the coverslip, without pushing on it in order to avoid squashing of the embryos

4. Breath out all the embryos, trying to put them in line in the delimited area of the slide (*see* **Note 22** and Fig. 5d step 2).

5. Repeat the last two steps as many times as necessary, taking five embryos from the first group each time

6. In one hand, with a very thin glass pipette aspirate as much solution around the embryos as possible. In the other hand, be ready, with the pipetman containing the antifading agent to put down a little drop of antifading agent on embryos immediately after the removal of the DNA staining solution (*see* **Note 23** and Fig. 5d step 3).

7. Repeat all the steps for the other groups of embryos.

8. Put a little drop of clear nail polish on each corner of the slide, next to the groups, in order to preserve the 3D-shape of the embryos (*see* Fig. 5d step 4).

9. Carefully put the cover slip on the slide helping yourself with a micro-spatula, starting from one side and getting down the other side slowly (*see* Fig. 5d step 5).

10. Secure with clear nail polish around the coverslip (*see* **Note 24** and Fig. 5d step 6).

11. Keep the slides at 4 °C until observation.

3.4 Microscope Observation

A wide-field microscope is not adequate to capture thick specimens like three-dimensionally 3D-preserved embryos, as it does not have enough z-axis resolution. Thus 3D-preserved embryos should be observed either on a fluorescent microscope equipped with structured illumination or on a confocal laser scanning microscope. It is also necessary to use a high magnification objective adapted to fluorescent observations (X63 Plan-Neofluar for example).

1. Place the slide on the microscope with the cover slip on the same side as the objective.

2. Locate the embryos with a low magnification objective thanks to the drawings performed on the back of the slide.

3. Use high-magnification objectives to observe the immunostaining. Oil-immersion ones are usually better and require a small drop of immersion oil that should be place between the cover slip and the objective.

4. Choose the fluorescence wavelength and the emission/excitation filters corresponding to the antibodies used during the immunostaining procedure (*see* **Note 25**).

5. Adjust the parameters of the microscope (acquisition time for example) in order to obtain the best signal-to-noise ratio: saturation of the detector should be avoided and the background noise level should be kept to a minimum.

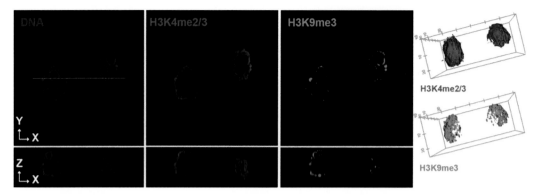

Fig. 6 Immunodetection of two epigenetic modifications in a two-cell mouse embryo produced by nuclear transfer. Two different epigenetic modifications (H3K4me2/3 in *red* and H3K9me3 in *green*) are immunostained. DNA is counterstained in *blue*. The figure presents confocal sections according to the classical *X/Y* axis but also according to the *X/Z* axis showing the depth of the nuclei, at the position indicated by the *white line*. The 3D-shape of the nuclei is clearly preserved. The corresponding 3D reconstructions obtained with the 3D Viewer Plugin from ImageJ are shown on the *right*. It shows that whereas H3K4me2/3 staining is present in the whole nuclei, H3K9me3 aggregates preferentially on one side of each nucleus

6. Perform Z-series acquisition with small z-steps in order to obtain the best 3D reconstruction afterwards (*see* **Note 26**). We also recommend taking 2–3 additional optical z-sections on top and below the limits of the staining (Fig. 6).

7. For comparison between groups and quantification purposes, the same objective, offset, amplifier gain, wavelength range, and dichromatic mirrors should be used for all embryos.

4   Notes

1. Ideally room temperature should be stable, between 22 and 25 °C, otherwise we recommend using a heating block set at 27 °C.

2. Aspirator tubes can be hand-made with HPLC clear tubing (diameter 2 mm) but are also commercially available (e.g. from Sigma-Aldrich). To this system you can add a filter tip and aspirate with mouth, or add a 1 or 2 ml syringe and control the valve with your hand.

3. To get embryo glass manipulation pipettes, stretch a glass capillary (or tip of a Pasteur pipette) in the flame in order to get very thin pipettes with diameters similar to the embryo ones. It is also easier to work with quite short pipettes (no more than 8 cm in total) and a little oblique.

4. Immunostaining is usually performed in commercially available glass dishes with one or several wells. Plastic 4-cell plates can also be used but are usually less convenient for small preimplantation embryos. The capacity of each well has to be ≥500 μl.

5. Primary antibodies are recognizing specific antigens and can be isolated from various species (mouse, rabbit, rat, human). Some primary antibodies can already be chemically linked to a fluorophore (direct fluorescent antibody). This decreases the number of steps in the staining procedure making the process faster and can reduce background signal by avoiding some issues with antibody cross-reactivity or nonspecificity. However, since the number of fluorescent molecules that can be bound to the primary antibody is limited, direct immunofluorescence is less sensitive. If the fluorescent signal is too weak, amplification of the signal may be necessary. Secondary components such as fluorescently-tagged secondary antibodies (indirect immunostaining) can then be used: multiple secondary antibodies will bind a single primary antibody and provide a more pronounced signal. This also allows more flexibility because different secondary antibodies can be used for a given primary antibody.

6. In mouse preimplantation embryos we usually start with overnight incubation at 4 °C but some antibodies may be incubated 1 h at 37 °C in an oven or a dry incubator.

7. An extensive assortment of nucleic acid stains is commercial available for DNA counterstaining of the embryos. Their fluorescence absorption and emission spectra span the visible-light spectrum from blue to near-infrared, making them compatible with many different types of secondary antibodies. We often use DAPI (blue), Yo-Pro-1 (green), Propidium Iodide (red), and Ethidium Homodimer 2 (red).

8. Glass cover slips are commercially available in a variety of widths, lengths, and thicknesses. We recommend square or rectangular ones, easier to place on the embryos, measuring 22×22 mm up to 22×40 mm. The thickness of the cover slip is crucially important for high-resolution microscopy. Most microscope objectives are designed for use with No. 1 cover slips (0.13–0.17 mm thickness). Use of cover slips that deviate from this thickness will result in spherical aberration and a reduction in resolution and image intensity.

9. A comparative experiment with a short fixation and a long one in parallel, using the same antibodies, should of course be performed to confirm that a long fixation step does not alter the immunostaining.

10. As embryos are known to be very sensitive to culture conditions and environmental changes, we advise to perform embryo fixation as quickly as possible and to transfer them directly from the collection/culture medium to the fixative solution, with no rinse step.

11. The global success of the immunostaining, especially the quality of the signal, resides in the perfect contact of the embryos with each incubation solution. It is thus necessary at each step

to properly transfer the embryos and move them in the glass dish as described (Fig. 3).

12. In some cases, when the immunodetection is not giving a good signal, permeabilization may be either (1) extended to 1 h, (2) performed with 1 % Triton X-100, (3) performed at higher temperature, e.g. 37 °C or (4) after removal of the zona pellucida by HCl or pronase treatments. This last step is very useful is other species than mouse.

13. If necessary you can add, after the saturation of nonspecific sites, a step of denaturation with HCl (2 N HCl at 37 °C for 1 h) in order to increase DNA accessibility to the primary antibodies. This is particularly important for antibodies directed against 5-MeC or 5-hMeC [14].

14. Antibodies have to be prepared in sterile conditions, either under an appropriate hood or next to a flame. The dilution of the primary antibody will be adjusted in preliminary experiments. The best is to determine the lowest concentration that still provides an immunostaining signal, with a good signal-to-noise ratio. You may also combine several primary antibodies if they are not derived from the same species (to avoid cross-reactivity upon addition of the secondary antibodies) or if they are direct fluorescent antibodies. Combining several primary antibodies however requires preliminary experiments with each individual antibody to control that they do not physically cross-react with each other. In order for example to combine 5-MeC or 5-hMeC immunodetection with the detection of modified histones we recommend to proceed in several steps: (1) incubate with the primary antibody directed against the modified histone(s), (2) wash with 0.05 % Tween-20 in PBS for 30 min, (3) post-fix in 4 % PFA for 25 min, (4) treated with HCl as recommended in **Note 13** and (5) incubate with the 5-MeC or 5-hMeC primary antibody.

15. If the immunostaining results are not satisfactory, showing too much background, especially big fluorescent dots due to antibodies aggregates, we recommend centrifuging the antibody (10 min at $12,000 \times g$ with a microcentrifuge) after the dilution, before taking the supernatant to prepare the drops.

16. Choice of the incubation length and temperature will be adjusted in preliminary experiments in order to find the conditions that provide the best signal-to-noise ratio.

17. 0.05 % of Tween 20 may be added in these washing steps in order to reduce the background signal if necessary.

18. If you want to work on several proteins and use several antibodies simultaneously, make sure that (1) primary antibodies are produced in different species, (2) each corresponding secondary antibody is coupled to a different fluorophore and

(3) that each chosen fluorophore can be distinguished by the microscope that will be used. Secondary antibody dilution will be selected according to the manufacturer's recommendations. We commonly use 1/200 or 1/300 dilutions with antibodies from Jackson ImmunoResearch.

19. The "post-fixation" is used to preserve the immunostaining when immediate observation on the microscope is not possible.

20. DNA stains concentrations: 1 μg/ml DAPI; 0.002 mM Ethidium Homodimer-2; 1 μg/ml Propidium Iodide.

21. Do not use the extremity of the slide: when using an inverted microscope for observations the objective could be damaged.

22. Each step has to be done carefully in order to preserve the 3D-shape of the samples.

23. Be particularly careful to not dry the embryos.

24. Do not push on the coverslip in order to avoid squashing.

25. Fluorescence wavelength and emission/excitation filters can be found on the antibodies manufacturers' websites. We recommend the Fluorescence SpectraViewer from Invitrogen, Life Technologies that allows all kinds of combinations and comparisons.

26. The minimal *z*-step will depend on the resolution in the depth direction of your microscope that is determined by the numerical aperture of the objective, the refractive index of the objective immersion media and the wavelength of the light used.

Acknowledgement

All the present and past members from the lab should be acknowledged for their hard work, especially Lydia Ruddick and Bénédicte Sanseau who corrected this book chapter. We are also grateful to Pierre Adenot and Renaud Fleurot for Confocal and Apotome microscopy on the MIMA2 platform (Microscopie et Imagerie des Microorganismes, Animaux et aliments). Work in the lab is supported by the REVIVE Labex.

References

1. Maalouf WE, Liu Z, Brochard V et al (2009) Trichostatin A treatment of cloned mouse embryos improves constitutive heterochromatin remodeling as well as developmental potential to term. BMC Dev Biol 9:11

2. Le Bourhis D, Beaujean N, Ruffini S et al (2010) Nuclear remodeling in bovine somatic cell nuclear transfer embryos using MG132-treated recipient oocytes. Cell Reprogram 12:729–738

3. Liu Z, Wan H, Wang E et al (2012) Induced pluripotent stem-induced cells show better constitutive heterochromatin remodeling and developmental potential after nuclear transfer than their parental cells. Stem Cells Dev 21:3001–3009

4. Yang CX, Liu Z, Fleurot R et al (2013) Heterochromatin reprogramming in rabbit embryos after fertilization, intra-, and interspecies SCNT correlates with preimplantation development. Reproduction 145:149–159

5. Schneider R, Grosschedl R (2007) Dynamics and interplay of nuclear architecture, genome organization, and gene expression. Genes Dev 21:3027–3043

6. Bernstein BE, Meissner A, Lander ES (2007) The mammalian epigenome. Cell 128:669–681

7. Mason K, Liu Z, Aguirre-Lavin T et al (2012) Chromatin and epigenetic modifications during early mammalian. Anim Reprod Sci 134:45–55

8. Aguirre-Lavin T, Adenot P, Bonnet-Garnier A et al (2012) 3D-FISH analysis of embryonic nuclei in mouse highlights several abrupt changes of nuclear organization during preimplantation development. BMC Dev Biol 12:30

9. Andrey P, Kiêu K, Kress C et al (2010) Statistical analysis of 3D images detects regular spatial distributions of centromeres and chromocenters in animal and plant nuclei. PLoS Comput Biol 6:e1000853

10. Beaujean N, Taylor J, Gardner J et al (2004) Effect of limited DNA methylation reprogramming in the normal sheep embryo on somatic cell nuclear transfer. Biol Reprod 71:185–193

11. Yang J, Yang S, Beaujean N et al (2007) Epigenetic marks in cloned rhesus monkey embryos: comparison with counterparts produced in vitro. Biol Reprod 76:36–42

12. Pichugin A, Le Bourhis D, Adenot P et al (2010) Dynamics of constitutive heterochromatin: two contrasted kinetics of genome restructuring in early cloned bovine embryos. Reproduction 139:129–137

13. Liu Z, Hai T, Dai X et al (2012) Early patterning of cloned mouse embryos contributes to post-implantation development. Dev Biol 368:304–311

14. Salvaing J, Aguirre-Lavin T, Boulesteix C et al (2012) 5-Methylcytosine and 5-hydroxymethylcytosine spatiotemporal profiles in the mouse zygote. PLoS One 7:e38156

Chapter 10

Visualization of Epigenetic Modifications in Preimplantation Embryos

Hiroshi Kimura and Kazuo Yamagata

Abstract

Epigenetic modifications such as DNA methylation and posttranslational histone modifications change drastically during embryonic development. To visualize histone modifications in living embryos, a Fab-based live endogenous modification labeling (FabLEM) technique has been developed. Here we describe the methods required for FabLEM experiments, including Fab preparation from IgG, its conjugation with a fluorescent dye, loading into cultured cells or mouse embryos, and imaging.

Key words Antibody, Epigenetics, Fab, Histone modification, Live-cell imaging, Preimplantation embryo

1 Introduction

During development and differentiation of mammalian embryos, epigenetic information, including DNA methylation and posttranslational modifications of histones, changes drastically [1–3]. After fertilization, paternal DNA is rapidly demethylated and is assembled into nucleosomes with maternally supplied histones, resulting in resetting most epigenetic marks. Histones in maternal chromatin are inherited from oocytes and maintain some posttranslational modifications during the early cell divisions. These parental-specific epigenetic modifications become largely indistinguishable after several cell divisions. During embryonic development, DNA methylation and histone modifications further change globally and locally depending on the developmental stages and cell types. So far these global alterations of epigenetic modifications in preimplantation and early embryos have been analyzed mostly by immunostaining using fixed samples with modification specific antibodies. Although fixed-cell-based analysis provides "snapshot" images in individual cells during different developmental stages, the behavior of epigenetic modifications in single embryos cannot be tracked. To overcome the limitation of fixed-cell-based analysis, live cell imaging

Nathalie Beaujean et al. (eds.), *Nuclear Reprogramming: Methods and Protocols*, Methods in Molecular Biology, vol. 1222, DOI 10.1007/978-1-4939-1594-1_10, © Springer Science+Business Media New York 2015

techniques using specific probes for epigenetic modifications have been developed [4]. DNA methylation has been visualized by using a methyl-DNA binding domain fused with a green fluorescent protein (MBD-GFP) [5]. To track histone modifications in living cells, we have developed a Fab (antigen binding fragment)-based live endogenous modification labeling (FabLEM) technique using fluorescently labeled Fabs, which are prepared from mouse monoclonal antibodies directed against specific modifications such as histone H3 acetylation, methylation, and phosphorylation [6, 7]. These methods have been applied for analyzing dynamic changes of epigenetic modifications in mouse preimplantation embryos. Indeed, different levels of histone H3 lysine 27 acetylation have been observed in embryos generated by in vitro fertilization and somatic cell nuclear transfer [7].

As the method to track DNA methylation in mouse embryos has previously been described elsewhere in detail [8], we here describe the protocol to visualize histone modifications in living mouse preimplantation embryos by FabLEM. The scheme is illustrated in Fig. 1. To perform FabLEM, Fab needs to be prepared from IgG using protease digestion, as the whole IgG molecules are generally not adequate for imaging histone modifications. This is because, first, whole IgG is too big (~150 kDa) to pass through the nuclear pore by free diffusion so cytoplasmically injected molecules cannot enter the nucleus until the nuclear membrane breaks down during mitosis. Second, whole IgG binds bivalently to target epitopes and this strong binding affinity may lead to long binding times that interfere with biological processes. In contrast, because

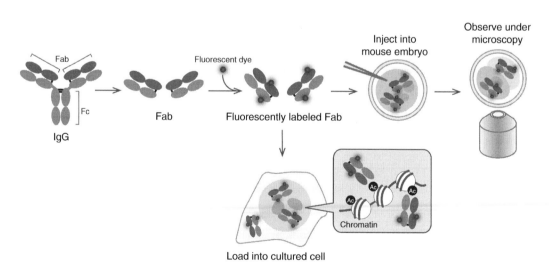

Fig. 1 Schematic illustration of FabLEM. From an antibody (IgG) that specifically recognizes a posttranslational modification on a specific residue, antigen binding fragment (Fab) is prepared, which is then conjugated with a fluorescent dye. The resulting fluorescently labeled Fab is now ready to be injected into cultured cells and mouse fertilized eggs for imaging

of its small size (~50 kDa), cytoplasmically injected Fab can enter the nucleus to bind its target modification and unbound molecules can diffuse out to the cytoplasm. Furthermore, the binding residence time of Fab ranges from sub-seconds to tens of seconds in living cells, which is a short enough time that endogenous proteins that may compete with Fab for modification targets can still access their sites with little interference [7]. Due to this binding kinetics, Fab injection and imaging does not affect cell growth and embryo development [6, 7].

2   Materials

Use Milli-Q ultrapure water for making all solutions.

2.1   Fab Preparation

1. Antibody: Prepare purified antibody using a protein A-column for monoclonal antibody or an antigen-affinity column for polyclonal antibody (*see* **Note 1**). 0.5–4 mg is generally used for Fab preparation (*see* **Note 2**). The concentration can be estimated by measuring the absorbance at *280* nm (Abs_{280}) with the extinction coefficient $210,000$ M^{-1} cm^{-1} and molecular mass of $150,000$. In the case of IgG, 1 unit of Abs_{280} corresponds to approximately 0.7 mg/mL (4.7 μM). If the subclass of IgG is uncertain determine the isotype using a kit (e.g., Roche; IsoStrip Mouse Monoclonal Antibody Isotyping Kit; 1493027).

2. A general buffer for antibody storage is phosphate buffered saline (PBS): To prepare 1 L PBS, dissolve 8 g NaCl (137 mM), 0.2 g KCl (2.7 mM), 1.15 g Na_2HPO_4 (anhydrous; 8.1 mM), and 0.2 g KH_2PO_4 (anhydrous; 1.5 mM) into 1 L water.

3. Protease-agarose beads and digestion buffer: It is more convenient to use an agarose bead-conjugated form of protease than the soluble form, as agarose beads can be readily separated from digested IgG (*see* **Note 3**). Fab is generally prepared by Papain digestion; however, as mouse IgG_1 is less susceptible to Papain and more digestable with Ficin [9, 10], Ficin is used for Fab preparation from mouse IgG_1.

   – For mouse IgG_1: Ficin-agarose beads (Thermo Scientific).

   – Ficin Digestion buffer: 25 mM L-cysteine, 5 mM EDTA (Ethylenediaminetetraacetic acid), 0.1 M sodium citrate, pH 6.0. To prepare 50 mL Ficin Digestion buffer, dissolve 220 mg L-cysteine·HCl·monohydrate in 50 mL of 0.1 M sodium citrate (pH 6.0) buffer containing 5 mM EDTA (*see* **Note 4**).

   – IgG_1 Binding buffer: 0.2 M sodium phosphate, 3 M NaCl. To prepare 100 mL IgG_1 Binding buffer, dissolve 2.8 g

Na_2HPO_4 (anhydrous) and 17.5 g NaCl into 100 mL water (*see* **Note 5**). The pH will be at ~8.2 when dissolved and can be used without adjusting pH. If the pH is lower than 8.2, raise the pH using 1 M NaOH. If the pH is higher (up to 9.0), the buffer can still be used as it is.

- For other mouse IgG subclasses or rabbit IgG use Papain-agarose beads (Thermo Scientific).

- Papain Digestion buffer: 20 mM cysteine, 10 mM EDTA, 20 mM sodium phosphate, pH 7.0. To prepare 50 mL Papain Digestion buffer, dissolve 35.1 mg cysteine ·HCl in 0.1 M sodium phosphate (pH 6.0) buffer containing 10 mM EDTA (*see* **Note 4**).

4. Glycine Elution buffer: 0.1 M Glycine·HCl, pH 2.8. To prepare 100 mL solution, dissolve 0.75 g glycine into 80 mL water, adjust the pH to 2.8 by adding conc. HCl, and add water up to 100 mL.

5. 1.5 M Tris–HCl, pH 8.8. To prepare 1 L solution, dissolve 121 g Tris base (tris-hydroxymethylaminomethan) into 800 mL water, adjust pH to 8.8 by adding conc. HCl, add water up to 1 L.

6. Protein A-conjugated beads: Protein A-Sepharose (GE Healthcare) or equivalent.

7. 2× SDS-gel loading buffer: 4 % sodium dodecylsulphate, 20 % glycerol; 100 mM Tris–HCl pH 6.8, 200 mM DTT.

8. pH indicator strips.

9. Centrifugal Filter Units: Amicon Ultra-15 50 K (50 k-cut off; Millipore; UFC905008; for concentrating IgG) and Amicon Ultra-15 10 K (10 k-cut off; Millipore; UFC901008; for concentrating Fab) (*see* **Note 6**).

10. Empty spin columns: Harvard Apparatus; Empty Macro SpinColumns; 743800.

11. Empty gravity flow columns: Polyprep columns; Bio-Rad; 731-1550.

12. 1.5 mL micro test tubes.

13. 2.0 mL micro test tubes.

14. 15 mL polypropylene test tubes.

15. A refrigerated microcentrifuge for spinning 1.5 mL tubes.

16. A refrigerated centrifuge for spinning 50 mL tubes.

17. SDS-polyacrylamide gel electrophoresis (SDS-PAGE) apparatus.

2.2 Fluorescent Dye-Conjugation

1. Dimethyl sulfoxide (DMSO) (*see* **Note 7**).

2. 1 M sodium bicarbonate, pH 8.3: Dissolve 830 mg $NaHCO_3$ in 10 mL water. This should yield the buffer pH at ~8.3. Tightly close the lid and store at 4 °C.

3. Amine-reactive fluorescent dyes (*see* **Note 8**): Alexa Fluor 488 SDP star (1 mg; Life Technologies) (*see* **Note 9**), Cy3 monoreactive dye (to label 1 mg protein; GE Healthcare), Cy5 monoreactive dye (to label 1 mg protein; GE Healthcare).

4. PBS (*see* above).

5. Desalting columns: GE Healthcare; PD MiniTrap G25.

6. Centrifugal Filter Units: Amicon Ultra-0.5 10 K (10 k cut-off; Millipore; UFC501024).

7. 1.5 mL micro test tube.

8. A refrigerated microcentrifuge for spinning 1.5 mL tubes.

2.3 Checking Fab Binding Activity Using Culture Cells

1. Cells and medium: Standard cells like human HeLa and mouse A9 can be used (*see* **Note 10**). These cells are routinely grown in Dulbecco's modified Eagle's medium (DMEM) containing antibiotics (100 U/mL streptomycin and 100 µg/mL penicillin) and 10 % fetal calf serum (FCS). For live cell imaging, use Phenol-red free DMEM containing the supplements to reduce background fluorescence.

2. Glass bottom dish: MatTek; P35G-1.5-10-C.

3. Glass beads: ≤106 µm diameter (Sigma-Aldrich; G4649) (*see* **Note 11**). When handling, wear gloves and work under a hood to prevent spreading glass beads through the air. To wash and sterilize glass beads, soak them in 2 M NaOH for 2 h and mix gently using a shaker or rotator. For example, pour glass beads into a 50 mL Falcon tube up to ~10 mL and then add 40 mL 2 M NaOH. Wash glass beads extensively with distilled water until the pH becomes neutral (as beads sediment in water, just wait a few min and then decant the water away). Wash a few times with 100 % ethanol. Air-dry overnight at room temperature.

4. Inverted fluorescence microscope.

2.4 Microinjection of Fluorescently Labeled Fab in Fertilized Eggs

1. Mouse eggs: To collect mouse eggs, use 7–12 weeks old female B6D2F1 (BDF1) mice, a cross between female C57BL/6 and male DBA/2 mice. Superovulate by intraperitoneal injections of 5 IU pregnant mare serum gonadotropin (PMSG) and 5 IU human chorionic gonadotropin (hCG) (Teikoku Zoki) at 48 h intervals. Collect cumulus cell-intact oocytes 13–15 h after hCG injection and soak in 0.2 mL drops of TYH (Table 1; [11]) medium covered with mineral oil.

2. Mouse sperm: Collect sperm from the cauda epididymidis of BDF1 males (>12 weeks) and soak in 0.2 mL drops of TYH medium. Capacitate by incubation for 2 h at 37 °C under 5 % CO_2.

Table 1
Formulation of media used for in vitro fertilization and microinjection (grams)

	TYH	CZB	HEPES-CZB
NaCl	3.488	2.385	2.385
KCl	0.178	0.18	0.18
KH$_2$PO$_4$	0.081	0.08	0.08
MgSO$_4$·7H$_2$O	0.146	0.145	0.145
Sodium lactate (60 % syrup)	–	2.65 mL	2.65 mL
Sodium pyruvate[a]	0.055	0.015	0.015
Glucose	0.5	0.5	0.5
BSA[a]	2	2.5	–
Polyvinyl alcohol	–	–	0.05
EDTA·4Na	–	0.02	0.02
NaHCO$_3$	1.053	1.055	0.21
HEPES	–	–	2.6
CaCl$_2$·2H$_2$O[a,b]	0.126	0.125	0.125
Penicillin/streptomycin[c]	0.025/0.038	0.025/0.035	0.025/0.035
200 mM L-Glutamine sol.[a,c]	–	2.5 mL	2.5 mL
10 mg/mL Phenol red sol.[c,d]	0.1 mL	–	0.1 mL
1.0 N NaOH	–	–	3.75 mL
Up to by milli-Q water	500 mL	500 mL	500 mL

[a]Make a stock solution by dissolving reagents other than these chemicals into water. After filtration through a 0.22 μm filter, stored at 4 °C. These marked chemicals are added to the stock solutions just before use (use in 1–2 month)
[b]Prepare 100× concentrated solution by dissolving into water, filtrate through 0.22 μm filter and store at 4 °C
[c]Commercially available premade solution could be used
[d]As this chemical causes background fluorescence, omit from the medium for imaging

3. Fertilized eggs: Inseminate cumulus-intact oocytes with capacitated sperm (final concentration 50–100 sperm/μL). After 1.5 h incubation at 37 °C under 5 % CO$_2$, disperse cumulus cells by brief treatment with hyaluronidase (Sigma-Aldrich; Type-IS, 150 U/mL) and collect anapahase/telophase II stage oocytes.

4. Chatot–Ziomek–Bavister (CZB) medium (Table 1; [12]).

5. HEPES-CZB (Table 1) containing 12 % Polyvinylpyrrolidone (PVP-HEPES-CZB).

6. Injection system. An inverted microscope (Olympus; IX-71) attached with a micromanipulator system (Narishige; ON3) with a Piezo-drive microinjector (Prime Tech, LTD; PMM-150H).

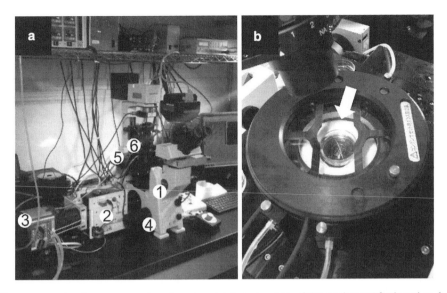

Fig. 2 Microscope setup for mouse embryo imaging. (**a**) A photograph of the equipment for imaging. A conventional inverted microscope (*1*) is attached to a Nipkow disk confocal unit (*2*), an EM-CCD camera (*3*), a *z*-axis motor (*4*), and an automatic *x–y* axis stage (*5*). For excitation, solid-state lasers (405, 488 and 561 nm) are used as light sources. Embryos are cultured in an incubator on the stage (*6*), which is set at 37 °C and gassed with 5 % CO_2 in air at 160 mL/min. (**b**) An incubation chamber on the stage. A glass-bottom dish (*arrow*) is put in the center of the chamber and surrounded by a water bath

2.5 Live-Cell Imaging of Preimplantation Development

1. An inverted microscope system (Fig. 2): e.g., Olympus; IX-71 equipped with a Nipkow disk confocal unit (Yokogawa Electric Corp; CSU series) (*see* **Note 12**), electron multiplying charge-coupled device (EM-CCD; Andor Technology; iXon series), Z motor (Ludl Electronic Products), auto X–Y stage (Sigma Koki), and incubator chamber (Tokai Hit; MI-IBC).

2. Laser light source: 488 and 561 nm lasers (CVI Melles Griot, Albuquerque, NM, USA).

3. Power meter: TB-200 (Yokogawa Electric).

3   Methods

3.1 Fab Preparation

1. It will take a whole day (~7–8 h) to complete Fab preparation (*see* Fig. 3 for the flow).

2. Prepare Digestion buffer just before use (*see* **Note 4**). If antibody is mouse IgG_1, use Ficin Digestion buffer. If antibody is not mouse IgG_1, use Papain Digestion buffer.

3. To exchange the buffer of IgG, pour desired amount of antibody (0.5–4 mg; up to 12 mL) into a centrifugal filter unit, Amicon Ultra-15 50 K (50 k-cut off) (Fig. 3; the first row) (*see* **Note 6**).

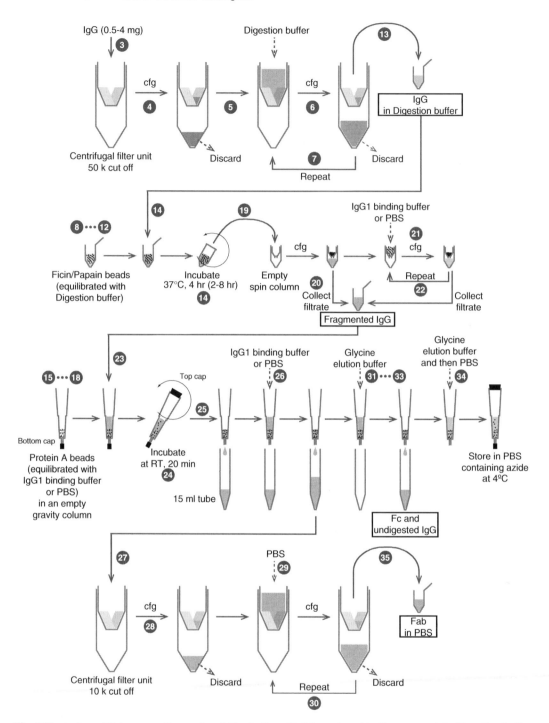

Fig. 3 Illustration of Fab preparation protocol. The buffer of IgG is replaced by the appropriate Digestion buffer, i.e., Ficin Digestion buffer for mouse IgG$_1$ or Papain Digestion buffer for other subclasses, using a centrifugal filter unit (50 k cut-off) by centrifugation (cfg) (the first row). IgG is mixed with appropriate protease-conjugated agarose beads, i.e., Ficin for mouse IgG$_1$ or Papain for other subclasses, to digest IgG into Fab and Fc (the second row). Fab is separated from Fc and undigested IgG by mixing with Protein A beads (the third row) and the buffer is replaced to PBS (the fourth row) using a centrifugal filter unit (10 k cut-off). The numbers in gray circles corresponds to steps in Subheading 3.1

4. Centrifuge the filter unit at $5,000 \times g$ for 15 min at 4 °C using an angle rotor (or at $4,000 \times g$ using a swing rotor). Antibody remains in the filter unit as typically <0.5 mL solution, but the volume depends on the initial volume and total amount of antibody. If the volume is larger than 0.5 mL, centrifuge a further 5–15 min to reduce the volume.

5. Throw away the filtrate at the bottom of the 50 mL tube, and pour Digestion buffer into the filer unit up to 12 mL.

6. Centrifuge the filter unit at $5,000 \times g$ for 15 min at 4 °C (or at $4,000 \times g$ using a swing rotor). If the volume is larger than 0.5 mL, centrifuge a further 5–15 min to reduce the volume.

7. Repeat **steps 5** and **6**.

8. During buffer replacement by centrifugation (**steps 5–7**), prepare protease-conjugated agarose (Fig. 3; the second row) (**steps 9–12**).

9. For moue IgG_1, aliquot 0.75 mL Ficin-conjugated agarose (33 % slurry; 0.25 mL settled resin) into a micro test tube. For other subclasses, aliquot 0.25 mL Papain-conjugated agarose (50 % slurry; 0.125 mL settled resin) into a micro test tube. Use a wide-bore or top-cut pipette tip to handle the resin.

10. Spin down the resin at $5,000 \times g$ for 1 min, and discard the supernatant.

11. Add 1 mL Digestion buffer to the resin (Ficin Digestion buffer for Ficin beads, or Papain Digestion buffer for Papain beads), mix by inverting, spin down ($5,000 \times g$ for 1 min), and discard the supernatant.

12. Repeat **steps 10** and **11** twice (total three times washing with Digestion buffer).

13. (From **step 7**) Transfer the IgG from the filter unit to a micro test tube. As IgG may sediment at the bottom, resuspend the solution well before the transfer. If the volume is <0.5 mL, rinse the filter unit using Digestion buffer. The final volume should be adjusted to approximately 0.5 mL. For SDS-PAGE analysis to validate Fab preparation, aliquot 5 µL into a micro test tube and store on ice (as "IgG" sample before digestion) (*see* Fig. 4 below).

14. Mix 0.5 mL IgG solution with the resin, and incubate the mixture at 37 °C for 4 h with rotation to digest IgG into Fab and Fc (*see* **Note 13**).

15. During this incubation, prepare protein A column (Fig. 3; the third row) (**steps 16–18**).

16. Pour Protein A beads slurry into a Polyprep column to settle 1 mL resin. Take bottom cap off and drain.

17. Pour 5 mL IgG_1 Binding buffer (for IgG_1 with Ficin beads) or PBS (for other subclasses with Papain beads) into the column.

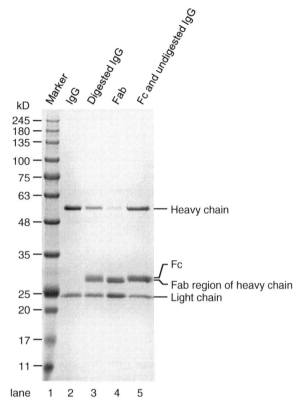

Fig. 4 SDS-PAGE to validate Fab preparation. An example of Fab preparation using mouse IgG_1 and Ficin-agarose beads is shown. Samples of IgG, digested IgG, Fab, and Fc plus undigested IgG were separated in a 10–20 % gradient SDS-polyacrylamide gel (Wako; SuperSep Ace) under reduced condition. Heavy and light chains are detected in the IgG sample (*lane 2*). Bands corresponding to Fc and Fab regions of the heavy chain (HC-Fab) appear at ~30 kDa after digestion with Ficin-agarose for 4 h (*lane 3*). Note that the size of HC-Fab varies depending on antibody; in this case, Fc and HC-Fab are not well separated due to their similar sizes. Judging from the band intensity of HC before and after Ficin digestion (*lanes 2* and *3*), approximately one-third of IgG appears to remain undigested. As no smaller band is seen, digestion time can be increased up to 6–8 h when preparing Fabs from the same antibody again. After incubation with Protein A beads, most Fc and undigested IgG were removed from the Fab fraction (*lane 4*). Fc and undigested IgG were eluted from Protein A beads (*lane 5*). Size standards (Wako; Wide-View III) are separated in parallel (*lane 1*)

18. Repeat **step 17**. Put the bottom cap back on.

19. (From **step 14**) To separate fragmented IgG from beads, transfer the mixture into an empty spin column settled onto a 2 mL micro test tube, and centrifuge at $5,000 \times g$ for 1 min.

20. Transfer the filtrate into a new 2 mL tube. Settle the column filter unit back to the original tube.

21. To wash out the residual fragmented IgG, add 0.5 mL IgG_1 Binding buffer (for Ficin beads), or PBS (for Papain beads), to the spin column containing beads (*see* **Note 5**). Spin down the column at $5,000 \times g$ for 1 min, and transfer the filtrate into the tube containing the first filtrate.

22. Repeat step 21 twice. Total 2 mL fragmented IgG should be yielded from the original and three washing filtrates. For SDS-PAGE analysis, aliquot 5 μL into a micro test tube and store on ice (as "fragmented IgG" sample).

23. Pour fragmented IgG (2 mL) into the bottom-capped column containing Protein A beads equilibrated with appropriate buffer (IgG_1 Binding buffer or PBS).

24. Close the top cap, mix the resin by inverting, and incubate for 20 min at room temperature with rotation. Fc and undigested IgG are captured by Protein A during this incubation, and Fab remains unbound in supernatant.

25. Remove the top cap, place the column above a 15 mL tube, and then remove the bottom cap to collect the unbound Fab fraction into the tube.

26. To recover the remaining Fab in the column, add 5 mL of the appropriate buffer (IgG_1 binding buffer or PBS) on top of the resin and collect the flow through in the 15 mL tube used in **step 25**.

27. To exchange the buffer and concentrate Fab, pour the flow through (total ~7 mL) into a centrifugal filter unit, Amicon Ultra-15 10 K (10 k cut-off) (Fig. 3; the fourth row) (*see* **Note 6**).

28. Centrifuge the filter unit at $5,000 \times g$ for 15 min at 4 °C. Throw away the filtrate at the bottom of the 50 mL tube.

29. Add PBS to the filter unit up to 12 mL and Centrifuge at $5,000 \times g$ for 15 min at 4 °C. Throw away the flow through.

30. Repeat three more times to replace the buffer to PBS and to concentrate Fab. The final volume should be <0.5 mL. If the volume is larger than 0.5 mL, extend the centrifugation time.

31. During the buffer exchange of Fab, elute Fc and undigested IgG from Protein A beads (Fig. 3; the third row, right side) (**steps 31–34**).

32. Set a new 15 mL tube containing 0.4 mL 1.5 M Tris–HCl (pH 8.8) for neutralization, under the column containing Protein A beads.

33. Apply 4 mL Glycine Elution buffer and collect the eluted fraction in the 15 mL tube. After mixing well, check the pH is ~8 using a pH indicator strip (*see* **Note 14**). If pH is lower, add more 1.5 M Tris–HCl (pH 8.8). This fraction is used as the "Fc" sample for SDS-PAGE analysis.

34. Wash the Protein A beads with 10 mL Glycine Elution buffer, and then with 15 mL PBS. Store in PBS containing 0.02 % sodium azide at 4 °C (*see* **Note 15**).

35. (from **step 30**) Transfer Fab from the filter unit to a micro test tube. As Fab may sediment at the bottom, resuspend the solution well by pipetting before the transfer. If the volume is <0.5 mL, rinse the filter unit using Digestion buffer. Adjust the final volume to approximately 0.5 mL.

36. Spin down briefly ($12,000 \times g$ for 5 min at 4 °C) to remove any insoluble materials and transfer the supernatant into a new micro test tube, to yield the "Fab" sample.

37. Measure the absorbance at 280 nm of "Fab" (**step 35**) and other samples (i.e., "IgG," **step 13**; "fragmented IgG," **step 22**; and "Fc," **step 33**) to estimate their concentration. 1 U of Abs_{280} corresponds to approximately 0.7 mg/mL. To estimate molar concentration of Fab, use an extinction coefficient of $70,000 \ M^{-1} \ cm^{-1}$ and a molecular mass of 50,000 (one-third of each of IgG). 1 U of Abs_{280} 0.7 mg/mL corresponds to approximately 14 µM. If the concentration of Fab is less than 1 mg/mL, centrifuge using the filter unit again to concentrate. The concentrated Fab sample can be stored at 4 °C or –80° in aliquots.

38. Analyze the samples (0.5–1 µg of each) by SDS-PAGE (Fig. 4) (*see* **Note 16**).

3.2 Fluorescent Dye Conjugation

1. It will take ~2 h to complete dye conjugation.

2. Dissolve amine-reactive fluorescent dye in 100 µL DMSO. Store at –20 °C (*see* **Note 17**).

3. Aliquot 100 µg of Fab and mix with PBS up to 90 µL, and then add 10 µL of 1 M sodium bicarbonate (pH 8.3). This yields 1 mg/mL Fab in 0.1 M sodium bicarbonate (pH 8.3) (*see* **Note 18**).

4. Add 1–4 µL of an amine reactive dye to the Fab solution and quickly mix well (*see* **Note 19**).

5. Incubate at room temperature with gentle mixing by rotation for 1 h. Protect from the light.

6. During this incubation period, equilibrate a column (PD MiniTrap G25) with PBS by passing PBS through the column at least three times.

7. (from **step 5**) After 1 h incubation, pour the dye–protein mixture onto the column, and let it settle into the resin (Fig. 5).

8. Put 500 µL of PBS onto the column and take the bottom cap off to drain. Two fluorescent fractions will soon be seen.

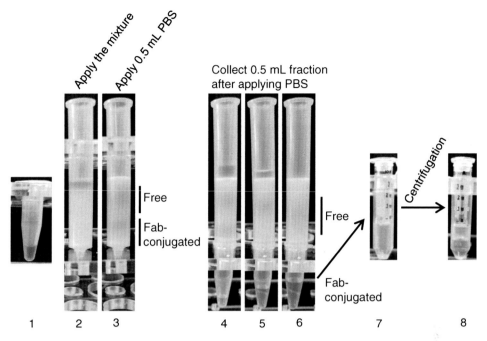

Fig. 5 Separation of fluorescent dye-conjugated Fab. Fab (100 μg) was incubated with Cy3 NHS ester in 100 μL reaction mixture for 1 h (*1*). After the mixture was applied to a desalting column (PD MiniTrap G25) (*2*), 500 μL of PBS was applied. Dyes conjugated to Fab migrated faster than free dyes (*3*). After setting the column on a micro test tube, 500 μL of PBS was applied to collect dye-conjugated Fab (*4–6*). The separation of dye-conjugated Fab from free dyes was clearly visible. In the case of Alexa Fluor 488, it may be difficult to see the fluorescence during column separation, but its fluorescence became clearer after the collection and concentration. Dye-conjugated Fab was transferred into a centrifugal filter unit (10 k cut-off) (*7*) and concentrated by centrifugation (*8*)

The first fluorescent fraction, which is dye-bound Fab, should reach close to the bottom of the column.

9. To collect the dye-conjugated Fab fraction, place the column in a new micro test tube and add 500 μL of PBS to elute. Do not collect the second fluorescence peak, which contains free dye molecules.

10. Transfer the eluted fraction into an Amicon Ultra-0.5 10 K (10 k cut-off), and centrifuge at $12,000 \times g$ at 4 °C for 10 min. This typically yields 50–80 μL solution in the filter unit.

11. Discard the filtrate at the bottom of the 2 mL tube, add 400 μL PBS to the filter unit, and repeat the centrifugation at $12,000 \times g$ at 4 °C for 10 min. This step is optional, but this helps to remove residual free dye molecules (*see* **Note 20**).

12. Suspend the solution in the filter unit by pipetting, as proteins may be sedimented at the bottom, and transfer the labeled Fab to a micro test tube.

Table 2
Summary of extinction coefficients and correction factors for each dye

Dye	Wavelength (nm)	Extinction coefficient [ε(dye)] (M^{-1} cm^{-1})	Correction factor for absorbance at 280 nm [CF]
Alexa488	488	71,000	0.11
Cy3	550	150,000	0.08
Cy5	642	250,000	0.05

13. Measure the absorbance at 280 and 494 (Alexa Fluor 488), 550 (Cy3), or 642 (Cy5) nm to calculate the protein concentration and dye:protein ratio.

$$\text{Dye concentration}\,(M) = \frac{\text{Abs(dye)}}{\varepsilon(\text{dye})}$$

$$\text{Fab concentration}\,(M) = \frac{\text{Abs}(280) - \text{CF(dye)} \times \text{Abs(dye)}}{\varepsilon(\text{Fab})}$$

$$\text{dye : protein ratio} = \frac{\text{Dye concentration}}{\text{Fab concentration}}$$

where the Abs(dye) is the absorbance for 1 cm path length at the wavelength suitable to each dye; ε(dye) is the extinction coefficient of each dye at the wavelength; Abs(280) is the absorbance for 1 cm path length at 280 nm; ε(Fab) is the extinction coefficient of Fab (70,000 M^{-1} cm^{-1}); and CF(dye) is the correction factor of each dye to compensate the absorption at 280 nm. To obtain Fab concentration as mg/mL, multiply the molar Fab concentration (M) by 50,000 (the molecular mass). The wavelength, extinction coefficient, and correction factor are summarized in Table 2.

14. Adjust the concentration of Fab to 0.5 or 1.0 mg/mL. Typical recovery rate is ~80 % of the starting material (i.e., ~80 μg Fab) and the dye:protein ratio is ~1–2.

3.3 Checking the Activity of Fluorescently Labeled Fab Using Cultured Cells

1. Plate cells on to a 35 mm glass-bottom dish and incubate at 37 °C in a CO_2 incubator overnight (see **Note 11**).

2. Take a glass-bottom dish out of a CO_2 incubator.

3. Remove the culture medium from the dish using an aspirator. Make sure to also remove residual media around the rim of the glass-bottom part of the dish (see **Note 21**).

4. Pipette 2–4 μL of dye-conjugated Fab (0.5 mg/mL) in the center of the glass coverslip region of the dish.

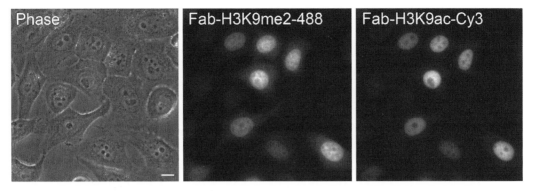

Fig. 6 Binding of fluorescently labeled Fabs to their targets in living cells. HeLa cells were loaded with Alexa Fluor 488-conjugated Fab specific to histone H3 dimethylated at lysine 9 (H3K9me2) and Cy3-conjugated Fab specific to histone H3 acetylated at lysine 9 (H3K9ac). Fab-H3K9me2 and Fab-H3K9ac exhibit slight-heterochromatic and euchromatic distributions, respectively, as observed by immunofluorescence using fixed cells, indicating that they can track the target modifications in living cells

5. Sprinkle dried glass beads onto the coverslip to distribute as a single layer or less. Avoid overlay (*see* **Note 22**).

6. After sprinkling beads, firmly strike the glass-bottom dish against the hood tabletop ~6–8 times. Each time raise the dish ~5–8 cm above the tabletop before striking down (*see* **Note 23**).

7. Immediately add 2 mL of prewarmed Phenol red-free medium to the dish.

8. Wash away glass beads by repeatedly washing with Phenol red-free medium.

9. Add 2 mL Phenol red-free medium to the dish and put it back into a CO_2 incubator or set it on a microscope for imaging. Incubate the cells at least 2 h before microscopic analysis. This allows Fab localization within the cell to reach a steady state.

10. Investigate the localization of Fab under a fluorescence microscope (Fig. 6).

3.4 Microinjection of Fluorescently Labeled Fab in Fertilized Eggs

1. Dilute fluorescently labeled Fab to 25–100 µg/mL using Milli-Q water (*see* **Note 24**).

2. Place an aliquot (0.5–1 µL) onto a micromanipulation chamber (Fig. 7a).

3. Place a 7 µL drop of HEPES-CZB medium and a 7 µL drop of PVP-HEPES-CZB in the same injection chamber (Fig. 7a).

4. Transfer anaphase II/telophase II oocytes to HEPES-CZB medium and inject Fab solution using a Piezo-drive microinjector with a narrow glass needle (1–3 µm diameter). Before Fab injection, suck PVP-HEPES-CZB once to wash and to prevent the stickiness of the inner wall of the glass needle. Once the Fab solution has been aspirated into the needle, apply

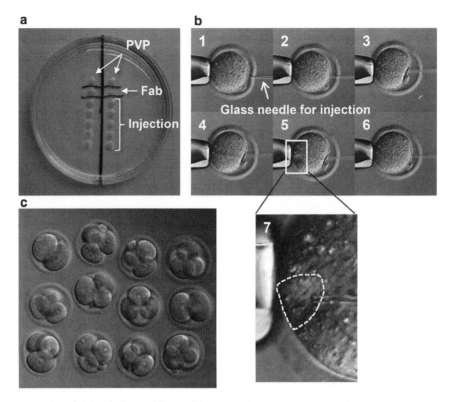

Fig. 7 Microinjection of Fab solution and live-cell imaging of preimplantation embryos. (**a**) A photograph of the injection chamber. A lid of a 60 mm dish is used. Seven microliter of PVP solution (12 % Polyvinylpyrrolidone diluted with HEPES-CZB medium) and HEPES-CZB medium for the injection, and >1 μL of Fab solution were placed on the chamber and covered by mineral oil. (**b**) Scheme of Fab injection using a piezo-drive microma-nipulator. Once Fab solution had been sucked into the glass needle (2–3 μm diameter), piezo pulses were applied to an oocyte placed in separate drops to break the zona pellucida (*panels 1* and *2*) and plasma mem-brane (*panels 3* and *4*). A few picoliters of solution were introduced into the oocyte (*panel 5*) and the needle was removed gently (*panels 6*). *Panel 7* is a magnified image of the boxed area of *panel 5* and shows the displace-ment of ooplasm by the emergence of the Fab solution from the tip of the needle (area indicated by dotted line). The volume of solution injected is controlled by seeing the extent of displacement. (**c**) Typical DIC images of embryos aligned in 3 × 4 formations for imaging using an UPlanApo 20× oil-immersion objective lens

piezo pulses to the oocyte to break the zona pellucida and oolemma. Introduce a few picoliters of solution into the oocyte and remove the needle quickly (Fig. 7b) (*see* **Note 24**).

5. Incubate Fab-injected embryos in CZB medium at 37 °C under 5 % CO_2 until imaging.

3.5 Live-Cell Imaging of Preimplantation Development

1. Set an inverted microscope with a heat-stage chamber to keep the temperature at the specimen at 37 °C and the air atmo-sphere at 5 % CO_2 (*see* **Note 25**).

2. Transfer Fab-injected embryos into a 5 μL drop of CZB medium in a 35-mm glass-bottom dish (MatTek) and align them to 3×4 formation using a glass needle that can be manipulated by hands.

In this way, images of 12 embryos in 1 field can be captured (Fig. 7c) when a 20× objective is used.

3. Place the dish in a chamber set on a microscope (*see* Fig. 2b).

4. Incubate at least 30 min to warm up the dish to 37 °C and stabilize the whole system (*see* **Note 26**).

5. Collect fluorescence and bright-field images. Under appropriate conditions, 51 z-stacks with 2 μm intervals can be collected every 15 min over 3–4 days (60–70 h) (*see* **Note 27**).

4   Notes

1. In the case of polyclonal antibody, antigen-affinity purification is essential, since the specific moiety occupies only a part of total IgG.

2. Typical recovery efficiency is ~30–40 %; i.e., when starting with 1 mg IgG, ~300–400 μg Fab will be prepared.

3. For Fab preparation using protease-conjugated beads, convenient kits are available (Thermo Scientific).

4. Cysteine-containing buffer should be freshly prepared and used within a day.

5. Mouse IgG_1 binds to protein A beads only in the presence of >1 M NaCl at >pH 7. To efficiently capture mouse IgG_1, this condition is better than using protein G beads under physiological salt conditions such as PBS.

6. Amicon Ultra-15 10 K (10 k cut-off) can also be used for IgG concentration, although longer centrifugation time is required. For Fab, Amicon Ultra-15 10 K (10 k cut-off) must be used, as Fab goes through 50 k cut-off filter (Amicon Ultra-15 50 K).

7. Use freshly opened DMSO, if possible, as water in air may be dissolved in "old" DMSO. The presence of water in DMSO can lead to degradation of the reactive group during storage.

8. In principle, Fab can be labeled with any fluorescent dye coupled with a reactive group such as *N*-hydroxysuccinimide (NHS) ester. When choosing a fluorophore to conjugate with Fab, both the optical and chemical properties should be considered. The optical property includes the excitation and emission wavelength, brightness and photostablility. Some dyes with excellent brightness and photostability, may not be suitable for FabLEM, as their conjugation may impair Fab epitope binding activity and/or they may have affinity to cellular components on their own. In our experience so far, the best dyes adequate for FabLEM are Alexa Fluor 488, Cy3, and Cy5, for green (488 nm excitation), red (543 or 561 nm excitation), and far red (633 or 647 nm excitation) fluorescence, respectively.

9. Other forms of amine-reactive Alexa Fluor 488 ester, like NHS (1 mg; Life Technologies; A20000) and TFP (1 mg; Life Technologies; A30005), also work well.

10. Cells that strongly attach to coverslips are suitable for bead-loading to check the function of fluorescently labeled Fab in living cells.

11. Fab's binding activity to the specific target in living cells can be validated by loading Fab into cultured cells, before applying to mouse embryos. A general technique for loading proteins into living cells is microinjection, but this method is tedious, low throughput, and requires equipment. The bead-loading method presented here is a simple and convenient alternative [7, 13, 14].

12. It is recommended to use a spinning disc confocal microscope for live mouse embryo imaging. A wide-field microscope is not adequate to capture thick specimens like the whole embryo, as it does not have enough z-axis resolution. Using a conventional laser scanning confocal microscope allows optical sectioning, but is more toxic to cells because every single spot needs to be illuminated by a strong excitation laser to produce photons sufficient for detection by a photomultiplier within a few microseconds. Compared to the spot scanning system, a spinning disc-type confocal microscope with numerous pinholes is less toxic to cells, as the laser power delivered to single spots is much lower during an image acquisition time of a few hundred milliseconds by EM-CCD camera. In other words, to get such a short image acquisition time with a spot scanning system, one would have to image each pixel much faster, but this would require a higher laser power to get similar signal-to-noise. Thus, the time taken to collect a single high-quality image with minimal phototoxicity is much shorter with a spinning disc confocal system using an EM-CCD camera than with a spot scanning system using a photomultiplier. This is especially important when collecting many z-stack images for thick specimen.

13. Incubation time depends on antibody and its concentration. When the digestion efficiency is low, increase the incubation time up to 8 h. In the case of over-digestion, decrease the incubation time.

14. This fraction is usually thrown away after checking the Fab preparation by SDS-PAGE. However, if a substantial amount of undigested IgG remains, Fab preparation can be restarted using this fraction. In this case, the incubation time with protease resin should be increased.

15. The protein A column can be used at least ten times for absorbing Fc and undigested IgG, but it should not be used to purify IgG.

16. If a substantial amount of undigested IgG remains, the incubation time with protease resin should be increased. In contrast, if faster-migrating bands than expected from Fab (25–30 kDa for both heavy and light chains) appear, this may indicate unwanted digestion within the Fab region. Shortening the incubation time with protease resin helps to restrict the digestion to the hinge region. If there are protease-susceptible sites within the Fab region, however, it may be difficult to prepare functional Fab.

17. Long-term storage over several month and/or repeated freeze–thaws may reduce the labeling efficiency, because the amine-reactive group is easily degraded in water, which is present in air and may be absorbed into DMSO. When thawing, do not open the lid until the temperature of the tube containing the dye solution becomes room temperature.

18. To keep the buffer conditions the same among different samples, we use PBS rather than water to fill up the volume. Then, the buffer is a mixture of nine parts PBS and 1 part 0.1 M sodium bicarbonate (pH 8.3). When using Fabs dissolved into buffer containing primary and secondary amines (such as Tris), replace the buffer with one which does not contain amines like PBS using a desalting column or a centrifugal filter unit. When scaling up or down, keep the concentration of Fab and dye the same. For example, for 50 μg of Fab, make a reaction mixture in 50 μL with 0.5 μL Alexa Fluor 488 SDP star.

19. It is important to mix as quickly as possible, because the reaction starts immediately. To achieve 1:1 dye:Fab ratio, we typically use 1 μL for Alexa Fluor 488 SDP star or Alexa Fluor 488-NHS ester, 1.5 μL for Cy3-NHS ester, and 4 μL for Cy5-NHS ester. As the labeling efficiency, however, depends on antibody and conditions (e.g., the actual room temperature), optimize the conditions when necessary, The labeling efficiency becomes lower after long-term storage and/or repeated use; in this case, increase the amount of dye.

20. Check the color, or fluorescence, of the filtrate. If it still fluoresces, repeat this step several times until the color becomes clear.

21. Be careful not to dry up the cells on the coverslip, as only a small volume of medium/Fab solution remains. There is no incubation step throughout the procedure, so perform the whole process quickly (but do not rush).

22. There are several ways to sprinkle the beads. You can make a tube containing glass beads with a nylon mesh (100 μm mesh size) cover on the top. Align the tube so the top (i.e., the exit of beads) is just above the glass-bottom dish. Tap the tube to drop the glass beads. As there is air flow in a hood, dropping

beads far above the dish will miss the target. You can also make a dish with a nylon mesh-bottom. Flip it onto the glass-bottom dish and, if necessary, tap once to distribute the beads. Alternatively, you can sprinkle beads without a nylon mesh using a spatula or a P-1000 pipette, but care must be taken to sprinkle just a single layer.

23. The loading efficiency and cell survival rate are a trade-off. If you strike down harder or more, you can load more, but more cells will die or detach from the dish. Therefore, it is very important to determine the condition of loading suitable for the cell type used. Fully confluent cells tend to peal in mass, while isolated cells peal more easily than those with neighbors. Thus, a confluency of 80–90 % is ideal in our experience.

24. The volume of Fab solution introduced into the ooplasm should be controlled by eye; the ooplasm at the tip of the needle is pushed away a little by the emerging solution (Fig. 7b). The concentration of Fab is a critical factor not only for imaging but also for embryo development. Indeed, although a higher concentration yields higher intensity, an excess amount of Fab injected hampers the developmental capacity of embryos, probably because it interferes with the natural protein machinery. Therefore, we usually determine the optimal concentrations of Fab solution that do not interfere with embryonic development by transferring to pseudopregnant females. In general, up to 100 µg/mL Fab does not inhibit embryo development. Diethylpyrocarbonate (DEPC)-treated water should not be used for dilution, as this chemical is a broad inhibitor of any proteins including RNase by modifying their histidine residues and might affect embryonic viability if its activity persists in the solution. Also, PBS is not suitable for the dilution of antibody if you inject the mRNA simultaneously to express fluorescent proteins (like histone H2B-GFP), as the RNA is degraded in this pH immediately.

25. Maintaining the temperature at 37 °C is important to keep both the embryo condition good and the focus stable. A heating stage featuring a very narrow thermo-sensor (Tokai Hit; TSU) is helpful. Check the fluctuation of the temperature using a thermometer with a data logger (e.g., KN Laboratories; Thermochron).

26. If starting image capture without the incubation step, a drift in z axis may occur at the early time points of imaging. Autofocus systems, like ZDC (Olympus) and Perfect Focus (Nikon), can overcome this problem. An x–y-axis motored stage allows imaging multiple positions.

27. It is important to avoid phototoxicity by reducing the laser power to allow normal embryo development. We usually confirm the embryonic integrity by transferring the embryos to

pseudopregnant females after imaging. In our experience, embryos develop normally when laser powers of 488 and 561 nm are set at 0.1 mW at the tip of a 20× objective lens (Olympus; UPlanApo 20× oil) with the exposure time 100 ms each. When a 40× objective lens was used under the same conditions, the embryonic development was severely arrested. During a total of 70 h imaging (from 1-cell to blastocyst stage), embryos developed normally when the interval between imaging was set at 7.5 and 15 min, but not at 3.75 min.

Acknowledgement

We thank Yuko Sato, Timothy J. Stasevich, and Yoko Hayashi-Takanaka for comments, figures, and data. The authors' work was supported by grants-in-aid from the Ministry of Education, Culture, Sports, Science and Technology, Japan Science and Technology Agency Core Research for Evolutional Science and Technology, and New Energy and Industrial Technology Development Organization.

References

1. Santos F, Dean W (2004) Epigenetic reprogramming during early development in mammals. Reproduction 127:643–651

2. Morgan HD, Santos F, Green K, Dean W, Reik W (2005) Epigenetic reprogramming in mammals. Hum Mol Genet 14:R47–R58

3. Bogdanović O, van Heeringen SJ, Veenstra GJ (2012) The epigenome in early vertebrate development. Genesis 50:192–206

4. Kimura H, Hayashi-Takanaka Y, Yamagata K (2010) Visualization of DNA methylation and histone modifications in living cells. Curr Opin Cell Biol 22:412–418

5. Yamazaki T, Yamagata K, Baba T (2007) Time-lapse and retrospective analysis of DNA methylation in mouse preimplantation embryos by live cell imaging. Dev Biol 304:409–419

6. Hayashi-Takanaka Y, Yamagata K, Nozaki N, Kimura H (2009) Visualizing histone modifications in living cells: spatiotemporal dynamics of H3 phosphorylation during interphase. J Cell Biol 187:781–790

7. Hayashi-Takanaka Y, Yamagata K, Wakayama T, Stasevich TJ, Kainuma T, Tsurimoto T, Tachibana M, Shinkai Y, Kurumizaka H, Nozaki N, Kimura H (2011) Tracking epigenetic histone modifications in single cells using Fab-based live endogenous modification labeling. Nucleic Acids Res 39:6475–6488

8. Yamagata K (2010) DNA methylation profiling using live-cell imaging. Methods 52:259–266

9. Mariani M, Camagna M, Tarditi L, Seccaman E (1991) A new enzymatic method to obtain high-yield F(ab')2 suitable for clinical use from mouse IgG1. Mol Immunol 28:69–77

10. Adamczyk M, Gebler JC, Wu J (2000) Papain digestion of different mouse IgG subclasses as studied by electrospray mass spectrometry. J Immunol Methods 237:95–104

11. Toyoda Y, Yokoyama M, Hoshi T (1971) Studies on the fertilization of mouse egg in vitro. Jpn J Anim Reprod 16:147–151

12. Chatot CL, Lewis JL, Torres I, Ziomek CA (1990) Development of 1-cell embryos from different strains of mice in CZB medium. Biol Reprod 42:432–440

13. McNeil PL, Warder E (1987) Glass beads load macromolecules into living cells. J Cell Sci 88:669–678

14. Manders EM, Kimura H, Cook PR (1999) Direct imaging of DNA in living cells reveals the dynamics of chromosome formation. J Cell Biol 144:813–821

Chapter 11

Live Embryo Imaging to Follow Cell Cycle and Chromosomes Stability After Nuclear Transfer

Sebastian T. Balbach and Michele Boiani

Abstract

Nuclear transfer (NT) into mouse oocytes yields a transcriptionally and functionally heterogeneous population of cloned embryos. Most studies of NT embryos consider only embryos at predefined key stages (e.g., morula or blastocyst), that is, after the bulk of reprogramming has taken place. These retrospective approaches are of limited use to elucidate mechanisms of reprogramming and to predict developmental success. Observing cloned embryo development using live embryo cinematography has the potential to reveal otherwise undetectable embryo features. However, light exposure necessary for live cell cinematography is highly toxic to cloned embryos.

Here we describe a protocol for combined bright-field and fluorescence live-cell imaging of histone H2b-GFP expressing mouse embryos, to record cell divisions up to the blastocyst stage. This protocol, which can be adapted to observe other reporters such as Oct4-GFP or Nanog-GFP, allowed us to quantitatively analyze cleavage kinetics of cloned embryos.

Key words Nuclear reprogramming, Nuclear transfer, Time-lapse imaging, Molecular imaging, Fluorescence microscopy, Transgenic mice, H2b-GFP

1 Introduction

When browsing the table of contents of this book, a general tendency is striking: most studies of somatic cell nuclear transfer (NT) embryos target *fixed embryos* of *predefined key stages* (e.g., morula or blastocyst). While these approaches undoubtedly yield invaluable information, they have two limitations: Because the bulk of reprogramming has already taken place by the time of analysis, it is nearly impossible to assess the grounds of failure or success of an embryo *retrospectively*. And the destructive nature of analyzing fixed specimen limits prediction of developmental success *prospectively*.

To overcome these blind spots, we sought to analyze the cell cycle progression of NT embryos using live embryo cinematography. Cell cycle analysis using live cell imaging has the potential to unveil mechanisms of development and to make predictions about

Nathalie Beaujean et al. (eds.), *Nuclear Reprogramming: Methods and Protocols*, Methods in Molecular Biology, vol. 1222, DOI 10.1007/978-1-4939-1594-1_11, © Springer Science+Business Media New York 2015

developmental success. Establishment of totipotency after NT requires two key events: First, critical embryonic genes must be activated in less than 24 h (first wave of mouse embryonic genome activation) in order for the somatic nucleus to support cleavage when maternal transcripts disappear [1]. Second, the relatively long cell cycle of somatic nucleus donor cells [2] must be converted to the precisely timed embryonic cleavage program [3–5] characterized by rapid successions of DNA replication and mitotic divisions lacking notable G1 or G2 phases. NT embryos that fail to adapt to the embryonic cleavage regime may be selected against. Therefore, a systematic dissection of the first cell cycles of cloned mouse embryos could unveil important mechanisms related to somatic reprogramming and cell cycle regulation. Moreover, it has been reported that progression of human in vitro fertilized (IVF) embryos to the blastocyst stage can be predicted by determining blastomeres' cleavage timings [6]. We propose that in cloned mouse embryos, the timing of completion of certain cell cycles may allow to predict successful reprogramming.

However, the light exposure necessary for time-lapse cinematography is highly toxic to cloned embryos. For example, using protocols for time-lapse cinematography considered safe for mouse fertilized embryos [7], we observed two-cell stage arrest in embryos produced by somatic cell NT while embryos produced by intracytoplasmic sperm injection (ICSI) formed blastocysts. We devised a combined bright-field and fluorescence time-lapse cinematography protocol that improved survival of NT embryos. We used an interference bandpass filter (580/10 nm) for bright field to exclude harmful wavelengths, and employed a mouse line ubiquitously and constitutively expressing a histone H2b-GFP transgene (Fig. 1). With these tools we determined cell cycle lengths of the first four cell cycles of mouse embryos cloned from ovarian cumulus cells and control embryos fertilized by ICSI. From the springboard of these innovations we were able to correlate cell cycle kinetics with blastocyst formation, and to make predictions validated by gene expression analysis [8]. In addition, we tracked chromosome movement and nucleus formation by imaging the fluorescent histone H2b-GFP, which is incorporated into the chromatin.

2   Materials

2.1   Mice

1. Use 6–8-week-old B6C3F1 (C57Bl/6J × C3H/HeN) mice as oocyte donors. Do always keep control embryos cloned from B6C3F1 cumulus cells in the incubator and in the incubation chamber on the microscope.

2. Mice expressing histone H2b-linked GFP have been generated by Chizuko Tsurumi and Benoit Kanzler and have been

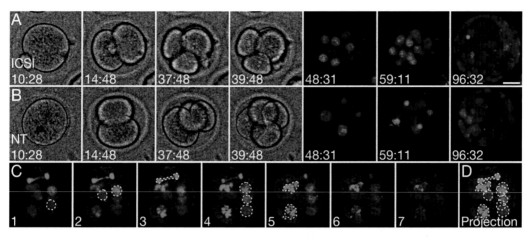

Fig. 1 Time-lapse cinematography of (**a**) ICSI and (**b**) NT mouse embryos. Until 48 hour post activation (hpa), bright-field images were captured every 20 min. From 48 until 96 hpa, confocal optical sections of H2b-GFP expressing embryos were captured every 20 min. Time-lapse movies were evaluated to obtain the timing of cleavages. *Numbers* indicate time (hh:mm) after activation. (**c**) Recording confocal sections allows distinguishing nuclei that could not be separated in conventional microscopy. (**d**) Maximum projection of confocal sections. *Dotted outlines* indicate nuclei or mitotic chromatin. Scale bar, 20 μm

published [8]. They can be obtained from the Mutant Mouse Regional Resource Centers (MMRRC) (Tg(HIST1H2B/GFP)#Mboi, MMRRC:036801).

3. Maintain mice in an environment with constant humidity, temperature, and photoperiod (e.g., 14L:10D) to minimize or prevent reported seasonal differences of embryo cleavage [9].

2.2 Collection of Mouse Oocytes and Nucleus Donor Cells

1. Pregnant mare's serum gonadotropin (PMSG).
2. Human chorionic gonadotropin (hCG).
3. Hyaluronidase (activity > 5,000 IU/mg).

2.3 Embryo Culture Media

1. Prepare culture media from stock solution on the day before the experiment. Sterilize by filtration through a cellulose acetate membrane (0.22 μm; preconditioned with medium). Equilibrate in humidified incubator at 37 °C under 5 % CO_2 in air for at least 12 h.

   – Normotonic (282 mOsm) HEPES-buffered CZB medium (HCZB), modified from the original recipe [10]: 81.62 mM NaCl, 4.83 mM KCl, 1.18 mM KH_2PO_4, 1.18 mM $MgSO_4 \cdot 7H_2O$, 5 mM $NaHCO_3$, 1.70 mM $CaCl_2 \cdot 2H_2O$, 31.30 mM sodium lactate, 0.27 mM sodium pyruvate, 0.11 mM disodium EDTA, 5.56 mM D-glucose, 20 mM HEPES, 5 mg/L phenol red. Prepare 4× stock solution in Milli-Q water and store refrigerated. Such concentrated stock is stable for at least 6 months in the dark at

4 °C. Add gentamicin sulfate (50 µg/mL) and polyvinyl-pyrrolidone (PVP, 40 kDa; 0.1 % w/v) shortly before medium use.

- Hypertonic (312 mOsm) HCZB: as above, but with 1 % w/v PVP and 90 % of the water volume.

- α-MEM (containing amino acids; 297 mOsm) supplemented with 2 mg/mL BSA (*see* **Note 1**) and 50 µg/mL gentamicin sulfate.

- Ca-free α-MEM modified from the Sigma recipe is described in Subheading 2.4.

2. Embryo-tested mineral oil. Extensively wash with Milli-Q water. Equilibrate in the incubator for at least 24 h but do not keep in the incubator for more than a week (in our experience it may become toxic). For long-term storage keep the oil in the cold and in the dark (e.g., inside refrigerator).

2.4 Micro-manipulation

1. Micromanipulation station at 28–30 °C room temperature fitted with a piezo actuator (e.g., PMM 150 FU, PrimeTech) and 40× DIC optics.

2. Cytochalasin B (Calbiochem) stock solution of 5 mg/mL in dimethylsulfoxide.

3. Glass-bottomed dish.

4. Micropipette pulled from borosilicate glass tubing, 12–15 µm in diameter, blunt end, backloaded with 2–3 µL of elemental mercury for spindle removal.

5. Micropipette as above, 8–10 µm in diameter, blunt end for nucleus transplantation.

6. Base activation medium is Ca-free α-MEM supplemented with 5 µg/mL cytochalasin B and 10 mM $SrCl_2$: 116.35 mM NaCl, 5.36 mM KCl, 1.02 mM NaH_2PO_4, 0.81 mM $MgSO_4 \cdot 7H_2O$, 26.19 mM $NaHCO_3$, 1.00 mM sodium pyruvate, 5.56 mM D-glucose, 1× MEM nonessential amino acid solution, 1× RPMI 1640 vitamins solution, 0.2 mg/L lipoic acid, 50 mg/L ascorbic acid-Na, 5 mg/L phenol red. Do not add essential amino acids normally contained in MEM. Prepare 4× stock solution in Milli-Q water and store refrigerated. Such concentrated stock is stable for at least 6 months in the dark at 4 °C. Add 2 mg/mL BSA (*see* **Note 1**) shortly before medium use. Add 5 µg/mL cytochalasin B and 10 mM $SrCl_2 \cdot 6H_2O$ when the medium is equilibrated for temperature and CO_2 (*see* **Note 2**). The complete activation medium (with supplements) has an osmolarity of 337 mOsm.

7. Four-well culture plates.

2.5 Live Embryo Microscopy Setup

1. *General considerations.* Careful design and critical optimization of all components of the imaging system is essential to successful live embryo imaging. In particular live cloned embryo imaging requires optimization of light paths because cloned embryos are much more sensitive to light than fertilized embryos. Since high image quality demands high light input, some image quality must be sacrificed in favor of lower light toxicity. In addition, any live cell imaging system must be physically stable so specimen do not move out of focus over the observation period. Embryos must be kept at stable temperature, humidity and atmosphere throughout imaging. Osmolarity and pH must remain constant over time. And last but not least, a fast and precise motorized microscope stage is needed to be able to image several conditions in one experiment. We used the microscope system described below that meets all the above criteria. For considerations on the microscopy lab, *see* **Note 3**.

   – PerkinElmer UltraVIEW RS3 Nipkow spinning disk confocal imaging system, fitted on a Nikon TE 2000U inverted microscope with a 20× objective lens (e.g., Nikon CFI Plan Fluor 20× MI—multi-immersion, N.A. 0.75; *see* **Note 4**).

   – Light source: For bright field, use a conventional halogen lamp (*see* **Note 5**). For fluorescence, use a laser source, for example, a three-line (488, 568 and 647 nm) Argon/Krypton laser (Melles Griot), to optimize fluorophore excitation and reduce undesired wavelengths.

   – A cooled interline CCD digital camera with high sensitivity (e.g., Hamamatsu ORCA ER) is required to allow high frame rates and reduced noise levels (*see* **Note 6**).

2. To access multiple positions in the incubation dish, use a motorized stage (e.g., Ludl BioPoint 2). A high resolution stepper motor for smooth and quiet movements with low vibration ensures that specimen stay at the same position throughout hundreds of rounds (*see* **Note 7**). To compensate unevenness and obliqueness of the vessel's floor and to allow fast focusing through the specimen, a piezoelectronic stepper motor moves the objective lens (Fig. 1c).

3. A mini incubator chamber (e.g., Tokai-Hit) maintains a constant gas phase (*see* **Note 8**) and humidity and keeps the dish and the environment at a constant temperature of 37 °C. A heated transparent lid prevents condensation of water droplets.

4. Use a high quality gas regulator to ensure a low but constant flow of CO_2–O_2 gas mixture (balanced with N_2). The gas

mixture should be humidified using a gas-washing bottle with a frit filled with pre-warmed sterile water (*see* **Note 9**).

5. 35 mm thin-bottom plastic dish (we use Greiner Bio-One, Lumox dish, hydrophilic) for embryo culture (*see* **Note 10**).

6. Protect the specimen from ambient light (*see* **Note 11**).

7. For image acquisition, we used the commercially available PerkinElmer Volocity software (*see* **Notes 12** and **13**).

8. For image analysis, we used ImageJ [11]. All data analysis and statistics were performed in R [12].

3   Methods

3.1   Collection of Mouse Oocytes and Nucleus Donor Cells

1. Stimulate oocyte and nucleus donor mice by injecting 7.5 IU of PMSG intraperitoneally in a carrier volume of 200–300 μL saline.

2. Superovulate mice 48 h later with 7.5 IU of hCG.

3. Collect cumulus-oocyte complexes 14–15 h post-hCG.

4. Place complexes into a 100 μL drop of hyaluronidase at 50 IU/mL in HCZB. Incubate for 20 min at room temperature (28–30 °C) to enzymatically remove cumulus cells.

5. Wash oocytes in normotonic HCZB and place in α-MEM until use.

6. Leave cumulus cells in hyaluronidase solution and store at 4 °C for up to 4 h until use.

7. Note that H2b-GFP is expressed very weakly in cumulus cells. An inexperienced experimenter may think that the PCR was false-positive.

3.2   Spindle Removal, NT, and Activation of Reconstructed Oocytes

For demanding experiments like live embryo imaging, only the best starting material should be used. Micromanipulations should be performed by skillful operators to ensure reproducibly high cleavage rates and sufficient numbers of embryos. Fertilized control embryos should be produced by ICSI, which underwent a similar amount of micromanipulation as embryos cloned by NT.

1. Perform all micromanipulations at 28–30 °C room temperatures. For a typical day's schedule *see* **Note 14**.

2. Place groups of 20 oocytes in 300 μL normotonic HCZB supplemented with 1 μg/mL cytochalasin B on a glass-bottom dish.

3. Remove the spindle-chromosome complex using the micropipette of 12–15 μm diameters.

4. Wash each group of oocytes in α-MEM, and store for 1 h in 500 μL of α-MEM medium in four-well plates at 37 °C under 5 % CO_2 in air.

5. Transplant cumulus cell nuclei into ooplasts 1–2 h later in hypertonic HCZB. Process groups of 30 oocytes for injection within a 10 min time span using the micropipette of 8–10 μm diameters.

6. Leave oocytes after NT for 5 min on the microscope stage to recover from possible mechanical trauma.

7. Gently transfer the nucleus-transplanted oocytes to a 1:1 mixture of HCZB and α-MEM and leave there for 1 h at room temperature (28–30 °C).

8. Activate reconstructed oocytes for 6 h in activation medium containing 5 μg/mL cytochalasin B at 37 °C.

9. Thoroughly wash activated embryos in α-MEM.

10. Culture embryos in groups of 100–150 in 500 μL α-MEM in four-well plates at 37 °C under 5 % CO_2 in air without oil overlay. Examine embryos at the four-cell stage 48 h after onset of culture, and supplement the medium with an additional 100 μL of α-MEM to ensure nutrition to the blastocyst stage.

3.3 Time-Lapse Cinematography

For preparations, *see* **Notes 15** and **16**.

1. When activation of reconstructed embryos is over, wash thoroughly and place them in groups of 20 close to each other into the drops on the imaging dish.

2. Carefully place the dish into the incubation chamber. Place a weight ring on top of the dish. Close the incubation chamber.

3. Move to the first imaging drop, focus on the embryos using dim red light. Save the imaging position in the software. Repeat with all imaging drops.

4. Capture bright field images for the first 48 h post activation using the settings described above (*see* **Note 17**). Check consistency of focus on a regular basis and correct it if necessary.

5. At 48 h post activation, record fluorescence images until 96 h post activation (Fig. 1).

3.4 Data Analysis and Statistics

1. Export time lapse movie. Prefer lossless file formats such as 16-bit TIFF. Those formats allow later adjustment of brightness/contrast to enhance details.

2. For every embryo, note the time of the first four cleavages of all cells, for example using ImageJ (*see* **Note 18**). Enter times into a spreadsheet, together with additional information such as blastocyst formation, mitosis errors and experimental details.

3. Calculate cell cycle lengths as difference between consecutive cleavages.

4. To test if the difference of certain cell cycle lengths between experiments is statistically significant, the Wilcoxon rank sum

test (in this case equivalent to the Mann–Whitney test) can be used. Use Fisher' exact test to evaluate if certain events (e.g., blastocyst formation, abnormal mitosis) occur at similar or different frequency amongst experimental groups.

4   Notes

1. Ensure to only use BSA with lowest endotoxin levels. Typically, clean BSA is very light in color; dark yellow BSA should be avoided. We have obtained good results using Millipore Probumin Media Grade.

   Presence of cytochalasin B, a mycotoxin that inhibits cytoplasmic division by blocking actin polymerization, is required to prevent extrusion of a polar body during oocyte activation. It is of utmost importance to store cytochalasin B stock solution at –80 °C. In our experience, its inhibitory effect on actin polymerization decays quickly if stored at –20 °C.

2. A precipitate of $SrCO_3$ will form after few hours if the base activation medium was not equilibrated at the time of $SrCl_2$ supplementation. Precipitation will lower concentration of $SrCl_2$ in the culture medium. Additionally, we think that the precipitate may damage the oolemma and hurt the embryo, in particular since cloned embryos have two holes in the zona pellucida.

3. An ideal site for long-term microscopic observations is a closed darkroom with little traffic. We recommend not to use neon lights or compact fluorescent lamps; these lights seem to emit toxic wave lengths. Use halogen bulbs instead. As vibration will cause embryos to move, there should be no devices causing vibrations like slamming doors, refrigerators, pumps, or centrifuges. The microscope should be placed on an anti-vibration table. Micromanipulation and microscopy stations may be situated in the same lab or in close by labs to prevent embryos from cooling down on transfer. Ideally, the room should be heated to 28–30 °C, as a compromise between the embryos' and operator's well-being.

4. Using a 20× objective lens, a field of view will fit about 12–15 embryos. Use a high transmission lens (characterized by a high numerical aperture), optimally using oil immersion. Avoid water or glycerol immersion.

5. In order to keep light emission stable, do not dim the lamp but use neutral density filters instead to decrease light intensity. To filter out toxic wavelengths and optimize the light spectrum for the sensitivity of the camera, use a bandpass filter (e.g., 580/10 nm). Use a shutter to turn on and off field illumination;

turning on and off the halogen bulb causes delays and unnecessarily exposes the embryos to light. To illuminate the field of view only, use a diaphragm.

6. Optimally, the camera should have a binning function to increase sensitivity when less detailed images are desired, for example when taking bright field images, and to increase resolution when more detailed images are desired.

7. To keep the dish in place, weight it down using a steel weight ring placed on top.

8. The gas mixture ratio consisting of 5 % CO_2, 5 % O_2, and 90 % N_2 is best kept constant by using pre-mixed gas sold in bottles.

9. Special attention has to be paid to hygiene of incubator and any tubing, for example by rinsing with 70 % ethanol before and after every usage, followed by thorough drying. Do not use disinfecting chemicals other than ethanol, as these may be embryo-toxic.

10. If using other dishes than the recommended ones, make sure they (a) are nontoxic to cloned embryos, (b) stand mineral oil for several days without melting, and (c) are suitable for imaging and nonfluorescent. The surface should not be too hydrophilic so drops of culture medium do not spread and merge.

11. Cloned embryos are extremely sensitive to light. Therefore careful protection against any ambient light is crucial. Shield windows completely against daylight. Only use red light if necessary. Use a metal plate with a central hole (diameter 3 cm) to cover the glass lid of the incubation chamber. Into the bottom of an empty pipette tip box, cut a hole large enough to fit the condenser of the microscope. Put the box upside down on top of the incubation chamber and enclose the condenser.

12. We used the following settings: until 48 h post activation, bright-field images were captured every 20 min, using a 4×4 binning and exposure time of 20 ms. After that until 96 h, confocal sections were captured in fluorescence mode 5 μm apart, using 488 nm laser with 0.5 mW laser power at the lens and an exposure time of 1 s. Choose to "Manage shutters for Maximum Sample Protection" in the software to close shutters when moving. These conditions allowed blastocyst formation of NT and ICSI embryos.

13. Use remote desktop connections to conveniently check your experiment on a regular basis from anywhere. Some commercially available software products even feature smartphone apps. Minor deviations of the focus or alike can be corrected online, and you are informed quickly if you have to return to the lab for major errors.

14. Oocyte collection, 7–8 a.m.; spindle removal, 8–10 a.m.; oocyte recovery from enucleation, 10–10:30 a.m.; nucleus

transplantation, 10:30 a.m.–1 p.m.; oocyte recovery from NT, 1–2 p.m.; oocyte activation 2–8 p.m.

15. At least 6 h before beginning the imaging session, fill the gas-washing bottle and the water reservoir of the incubation chamber with sterile water, and turn on gas flow and heat the incubation chamber to sufficiently equilibrate the system. Test all appliances. Set the focus using a mock dish so you don't have to expose valuable embryos to light unnecessarily for focusing.

16. Prepare the imaging dish: place 12 drops of 20 μL α-MEM each into the center of a thin-bottom dish plus one control drop close to the edge of the dish. Cover with equilibrated mineral oil. Equilibrate in the incubator.

17. Since H2b-GFP is very weakly expressed in cumulus cells, embryos reconstructed from transgenic cells are not fluorescent initially. GFP will first be visible at early four-cell stage, when enough protein has been assembled from the transplanted nucleus. If using hemizygous males for ICSI, only 50 % of the ICSI embryos will be fluorescent.

18. We have not found a program capable of automatizing this process.

Acknowledgements

The authors are thankful to Chizuko Tsurumi and Benoit Kanzler for making the H2b-GFP-transgenic mouse strain, without which this project would not have been possible. The strain is publicly available from The Mutant Mouse Regional Resource Center 8U42OD010924-13. We thank our coauthors of the original publication [8], Telma Esteves, Franchesca Houghton, Marcin Siatkowski, Martin Pfeiffer and Georg Fuellen for valuable input and the Max Planck Institute for Molecular Biomedicine, Münster, in particular Hans Schöler, for infrastructural support.

References

1. Vassena R, Han Z, Gao S, Latham KE (2007) Deficiency in recapitulation of stage-specific embryonic gene transcription in two-cell stage cloned mouse embryos. Mol Reprod Dev 74(12):1548–1556

2. Becker KA et al (2006) Self-renewal of human embryonic stem cells is supported by a shortened G1 cell cycle phase. J Cell Physiol 209(3):883–893

3. Day ML, Winston N, McConnell JL, Cook D, Johnson MH (2001) tiK+ toK+: an embryonic clock? Reprod Fertil Dev 13(1):69–79

4. Goval JJ, Alexandre H (2000) Effect of genistein on the temporal coordination of cleavage and compaction in mouse preimplantation embryos. Eur J Morphol 38(2):88–96

5. Zuccotti M et al (2002) Mouse Xist expression begins at zygotic genome activation and is timed by a zygotic clock. Mol Reprod Dev 61(1):14–20

6. Wong CC et al (2010) Non-invasive imaging of human embryos before embryonic genome activation predicts development to the blastocyst stage. Nat Biotechnol 28(10):1115–1121

7. Yamagata K, Suetsugu R, Wakayama T (2009) Long-term, six-dimensional live-cell imaging for the mouse preimplantation embryo that does not affect full-term development. J Reprod Dev 55(3):343–350

8. Balbach ST et al (2012) Nuclear reprogramming: kinetics of cell cycle and metabolic progression as determinants of success. PLoS One 7(4):e35322

9. Wang JP et al (1992) Seasonal variation in cell cycle during early development of the mouse embryo. J Reprod Fertil 94(2):431–436

10. Chatot CL, Ziomek CA, Bavister BD, Lewis JL, Torres I (1989) An improved culture medium supports development of random-bred 1-cell mouse embryos in vitro. J Reprod Fertil 86(2):679–688

11. Schneider CA, Rasband WS, Eliceiri KW (2012) NIH Image to ImageJ: 25 years of image analysis. Nat Methods 9(7):671–675

12. R Development Core Team (2011) R: a language and environment for statistical computing. R Foundation for Statistical Computing, Vienna, Austria

Chapter 12

Analysis of Nucleolar Morphology and Protein Localization as an Indicator of Nuclear Reprogramming

Olga Østrup, Hanne S. Pedersen, Hanne M. Holm, and Poul Hyttel

Abstract

When a cell is reprogrammed to a new phenotype, the nucleolus undergoes more or less dramatic modulations, which can be used as a marker for the occurrence of the reprogramming. This phenomenon is most pronounced when differentiated cells are reprogrammed to totipotency when they are submitted to cloning by somatic cell nuclear transfer. However, when cells are reprogrammed by less fundamental means, as for example treatment by Xenopus extract or expression of pluripotency genes, more subtle nucleolar modulations can also be noted. The monitoring and understanding of the reprogramming-related nucleolar modulations are based upon detailed knowledge about the nucleolar changes that occur during normal development from the developing oocyte over oocyte maturation and fertilization to the activation of the embryonic genome in the early embryo. Below, the ultrastructural and molecular modulations of the nucleolus are summarized in this developmental context, but also as they occur in assisted reproductive technologies such as in vitro fertilization and somatic cell nuclear transfer. Moreover, detailed protocols for monitoring the nucleolar changes by transmission electron microscopy and immunocytochemistry are presented.

Key words Nucleolus, Nucleus, Transmission electron microscopy, Immunocytochemistry, Assisted reproductive technology

1 Introduction

Cellular reprogramming represents a redirection of cell fate phenotype and consists of complex modulations in metabolic, proteomic, epigenomic, and genomic functions. The epigenomic and genomic modulations are spatially located to the nuclear compartment of the cell and are, thus, often referred to as nuclear reprogramming. Nuclear reprogramming is initiated at the molecular level (e.g., epigenetic changes in histones and DNA methylation or changes in DNA sequences by introduction of transgenes) and is consequently reflected in altered nuclear ultrastructure (e.g., chromatin and nucleolar morphology) which subsequently reflects in phenotypic alterations of cell morphology.

Nathalie Beaujean et al. (eds.), *Nuclear Reprogramming: Methods and Protocols*, Methods in Molecular Biology, vol. 1222, DOI 10.1007/978-1-4939-1594-1_12, © Springer Science+Business Media New York 2015

The nucleolus is the most prominent nuclear organelle, and in oocytes and embryos this nuclear compartment undergoes dramatic and well-described changes in relation to nuclear reprogramming. Hence, during normal development as well as at application of assisted reproductive technologies such as in vitro fertilization or somatic cell nuclear transfer, the nucleolus can be used as a morphologically visible and sensitive marker of nuclear reprogramming.

The mostly known nucleolar function is ribosome biogenesis including synthesis and processing of ribosomal RNA (rRNA). The RNA results from transcription of the rRNA genes (rDNA) in the nucleolus where the rRNA is subsequently processed and packed with proteins into the ribosomal subunits required for protein synthesis (for reviews, *see* refs. 1, 2). As visualized by transmission electron microscopy, the functional nucleolus is composed of three well-defined sub-compartments, namely the fibrillar centers (FCs), the dense fibrillar component (DFC), and the granular component (GC), and a nucleus presenting these features is referred to as being fibrillogranular (Fig. 1; for review, *see* ref. 3). The rDNA, the transcription factors (e.g., RNA polymerase I, upstream binding factor (UBF)) [4, 5], and the nascent rRNA are mainly located in the FCs and the inner portion of the DFC, together with proteins engaged in the early processing of the rRNA (e.g., fibrillarin). Proteins involved in the later steps of rRNA processing and ribosome subunit formation (e.g., nucleolin and nucleophosmin (B32)) are, in turn, located in the outer DFC and the GC.

Looked upon in a developmental perspective, the first ultrastructurally recognizable component related to assembly of the nucleolus appears in the pronuclei of the zygote in the form of nucleolus precursor bodies (NPBs) appearing as electron-dense compact spheres (for review, *see* ref. 6). The NPBs are more or less inherited from the oocyte. Hence, during the growth of the oocyte, a fibrillo-granular nucleolus sustains ribosome synthesis and translation, but at the end of the oocyte growth phase, the nucleolus is packed into an inactive nucleolar remnant (Fig. 2), which is dissolved at resumption of meiosis during oocyte maturation, and which emerges again in the zygote upon fertilization in the form of the NPBs (Fig. 3). The NPBs act as physical harbors for the development of functional nucleoli in the embryo, a process referred to as nucleologenesis (for review, *see* ref. 7). During this process, NPBs are gradually transformed and structures of FC, DFC and GC appear either on the rim of NPB (e.g., pig, mouse) or in vacuoles formed in the NPB (e.g., cattle) [8–10]. Nucleologenesis results in the formation of a functional fibrillogranular nucleolus and its timing coincides with the major embryonic genome activation (Fig. 3; [11, 12]). Hence, nucleolar transformation can be used as a marker of chromatin reprogramming. Moreover, recent investigations

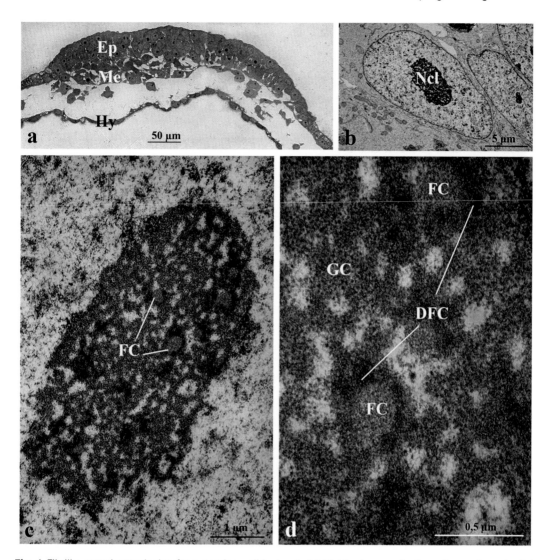

Fig. 1 Fibrillo-granular nucleolus from porcine epiblast cell. (**a**) Light micrograph of porcine embryonic disk from Day 11–12 embryo displaying the epiblast (Ep) in the process of gastrulation where mesoderm (Me) is ingressing into the space between the epiblast and the hypoblast (Hy). (**b**) Electron micrograph showing the nucleus of an epiblast cell with the nucleolus (Ncl) located in the center. (**c**) Electron micrograph of the nucleolus of an epiblast cell showing fibrillar centers (FC). High power electron micrograph of an epiblast cell nucleolus showing fibrillar centers (FC) surrounded by dense fibrillar component (DFC), and with the granular component (GC) forming a network in the organelle

have shown that the NPBs are also likely to serve as anchors of specific chromatin domains in early mammalian embryos [13–19]. During reprogramming, the ultrastructural and molecular changes related to the nucleolus are therefore of importance, not only for assessment of ribosomal gene remodelling but also as an indirect marker for assessment of reprogramming of other nuclear compartments.

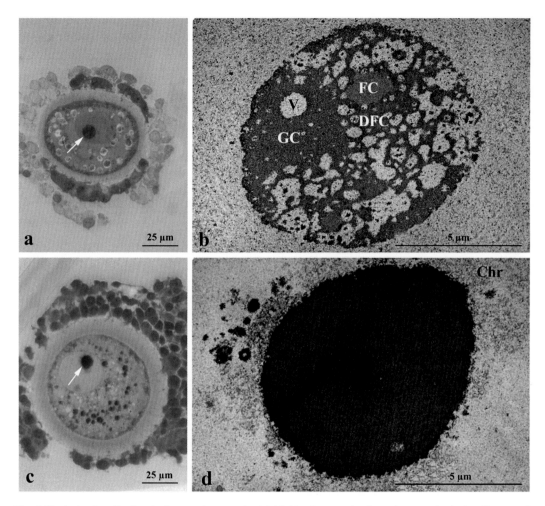

Fig. 2 Nucleolar inactivation in the porcine oocyte. (**a**) Light micrograph of porcine oocyte during the growth phase. Note the large reticulated nucleolus (*arrow*). (**b**) Electron micrograph of the nucleolus of the growth phase oocyte displaying a fibrillar center (FC), dense fibrillar component (DFC), and granular component (GC) as well as vacuoles (V). (**c**) Light micrograph of fully grown porcine oocyte. Note the compact inactive nucleolar remnant (*arrow*). (**d**) Electron micrograph of the nucleolar remnant of the fully grown oocyte associated with chromatin (Chr)

Nucleolar reprogramming has been previously successfully used for evaluation of developmental competence of embryos produced by various methods such as in vitro fertilization and cloning by somatic cell nuclear transfer [20–23]. However, over the past years new technologies have emerged with respect to reprogramming of somatic cell phenotype such as treatment with Xenopus extracts or introduction of key pluripotency genes or their products for generation of induced pluripotent stem cells (iPSCs; [24, 25]). During such somatic cell reprogramming the nucleolar compartment is not disassembled into NPBs as described for the embryonic reprogramming. Hence, monitoring somatic cell

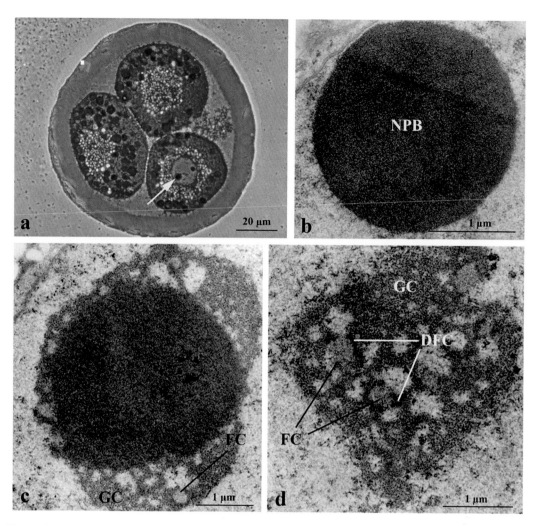

Fig. 3 External nucleologenesis in the 4-cell porcine embryo. (**a**) Light micrograph of early porcine 4-cell embryo. Note the nucleolus precursor body (*arrow*). (**b**) Electron micrograph of nucleolus precursor body (NPB) from early 4-cell porcine embryo. (**c**) Electron micrograph of external nucleologenesis in late 4-cell porcine embryo where a fibrillar center (FC) and granular component (GC) is deposited on the outside of the nucleolus precursor body. (**d**) Electron micrograph of fibrillo-granular nucleolus in late 4-cell porcine embryo displaying fibrillar centers (FC), dense fibrillar component (DFC), and granular component (GC). Partly from Hyttel et al. [8]

reprogramming poses new challenges with respect to the use of the nucleolar compartment as a marker. However, the underlying rRNA synthesis is tightly regulated in response to metabolic and environmental changes, and the ribosomal genes are epigenetically remodelled within a few hours after induction of differentiation, reprogramming, or stress [26–29]. During stress, energy dependent nucleolar silencing complex (eNoSC) targets rRNA promoters and inhibits excessive rRNA synthesis in order to redirect energetic resources towards elimination of stress factors [27].

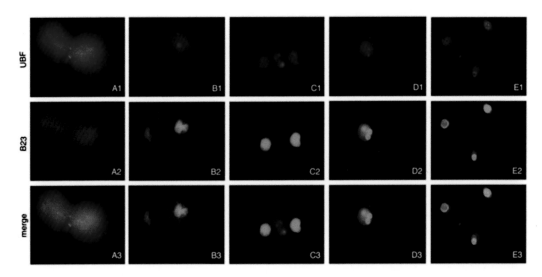

Fig. 4 Allocation of nucleolar proteins in early murine embryos during embryonic genome activation. Fluorescent micrographs of upstream binding factor (UBF; **a1–e1**) and nucleophosmin (B23; **a2–e2**) in murine embryos at early 2-cell stage (**a** and **b**), late 2-cell stage (**c** and **d**), and 4-cell stage (**e**), showing continuous allocation of these nucleolar proteins to forming nucleoli. In stage before the genome activation (**a** and **b**), UBF and B23 display dispersed cytoplasmic and later on also nucleoplasmic localization. During the genome activation (**c** and **d**), almost all proteins are transported from cytoplasm into nucleoplasm, and subsequently localized into forming nucleolar compartments. In embryos with active transcription (**e**), UBF and B23 display specific, distinct localization into foci (UBF) or nucleolar rings (B23). From [21]

Similarly, differentiation and related metabolic changes in the cell, also affect rRNA synthesis rate. However, during this process, active rRNA promoters are remodelled via nucleolar remodelling complex (NoRC), enabling long-term epigenetic silencing of numbers of rRNA genes so that rRNA production correlates with metabolic activity of particular cell type [28, 29].

Nucleolar reprogramming in early embryos and in somatic cells can be assessed on a basic level by two methods. Morphological changes are detectable by transmission electron microscopy. The more functional basis of nucleolar reprogramming can be captured by visualization of nucleolar proteins by immunocytochemistry. Immunocytochemistry is efficiently performed on whole mount embryos, which allows for example for double labelling of two different proteins (Fig. 4). The optimum would be to combine immunocytochemistry with transmission electron microscopy in the same specimen. However, the technology for ultrastructural immunocytochemistry is very time consuming, complex, and may not work for certain antibodies. Hence, a combination of electron microscopy performed on a subset of embryos and immunocytochemistry on another subset is often the method of choice.

Even though immunolabeling and electron microscopy are well-established methods, working with early embryos imposes several challenges for processing and evaluation. The present chapter will describe protocols for immunolabeling and processing for transmission electron microscopy that provide optimal conditions for assessment of nucleolar reprogramming in preimplantation embryos. The protocols are specified for evaluation of changes in ribosomal biogenesis during reprogramming. However, they are broadly applicable also for evaluation of other nucleolar functions and their changes triggered by reprogramming in somatic cells.

2 Materials

2.1 Embryo Fixation

1. Dissecting microscope with appropriate working distance.
2. 4-well dishes.
3. Glass capillaries for embryo transfer.
4. 10 mM phosphate buffered saline (PBS).
5. 4 % paraformaldehyde (PFA) in PBS.
6. Washing solution: PBS with 0.25 % bovine serum albumin (BSA) and 0.1 % polyvinyl alcohol (PVA) in PBS.
7. 3 % glutaraldehyde prepared from 25 % glutaraldehyde in 0.1 M sodium-phosphate buffer (Sorensen, *see* ref. 30).

2.2 Whole Embryo Immunolabeling

1. Shaker/rocker.
2. 96-well plate with rounded bottom.
3. Cover slips (24 × 24 mm).
4. Glass slides (normal).
5. Sheet reinforcements ring labels.
6. 0.1–1 % Triton X-100 in PBS for permeabilization.
7. 2 % BSA in PBS for blocking.
8. 0.25 % BSA/0.1 % Triton X-100 in PBS (washing solution).
9. Fluorescent mounting medium.
10. Hoechst (1 µg/ml) or DAPI (1 µg/ml).
11. Antibodies: mouse monoclonal anti-UBF, mouse monoclonal anti-B23, goat anti-mouse Alexa Fluor 488 and Alexa Fluor 594.

2.3 Embryo Processing for Transmission Electron Microscopy

1. 50 ml Erlenmeyer flask.
2. Microwave oven.
3. Aluminum foil.
4. Thermic heater adjustable to 50–60 °C.
5. Dissecting microscope.
6. Disposable pipettes (3 ml pipettes, standard).
7. Rotator (4 RPM).

8. Glass containers ($50 \times 19 \times 0.17$ mm) with cork stopper VWR SCERB40501000.

9. Grids for TEM: copper grids coated with 2 % collodion in amylacetate.

10. 0.1 M sodium-phosphate buffer.

11. 4 % agar in distilled water.

12. Distilled water.

13. 50, 70, 96 and 99 % ethanol.

14. Ultrapure ethanol.

15. Propylene oxide.

16. 1 % OsO_4 in 0.1 M sodium-phosphate buffer for post-fixation.

17. 1 % toluidine blue in 1 % sodium borate.

18. Nail varnish.

19. Liquid nitrogen.

20. 0.5 % uranyl acetate in water.

21. Epon resin.

22. Lead citrate [31].

23. Embedding molds.

3   Methods

3.1   Embryo Fixation

3.1.1   Embryo Fixation for Immunolabeling

1. Briefly rinse embryos two to three times in PBS to avoid the excess culture media containing serum (*see* **Note 1**).

2. Fix embryos for 20 min at room temperature (RT) with 0.5–1 ml of 4 % PFA (*see* **Note 2**).

3. Carefully transfer the embryos into washing solution and rinse 3×5 min under gentle agitation.

4. Fixed embryos can be stored up to 1 week at 4 °C in washing solution (preferred) or up to 2–3 months in 1 % PFA/PBS.

3.1.2   Embryo Fixation for Transmission Electron Microscopy

1. Briefly rinse embryos two to three times in PBS to avoid the excess culture media containing serum.

2. Fix embryos in 3 % glutaraldehyde (pH 7.2–7.4) for 1 h at 4 °C (*see* **Note 3**).

3. Rinse embryos two to three times in 0.1 M sodium-phosphate buffer to remove the excess glutaraldehyde.

4. Fixed embryos can be stored in 0.1 M sodium-phosphate buffer for 1 month at 4 °C, if properly covered with parafilm to avoid evaporation. Immediate processing is, however, recommended.

3.2 Immuno-detection of Nucleolar Proteins and Visualization of Nucleolar Ultrastructure

All steps are performed under gentle agitation ≤20 rpm. Embryos are placed one to six per well into 96-well plate with rounded bottoms and transferred between individual wells containing 100–200 µl of working solution. Due to the possible detachment of blastomeres in early embryos whose zona pellucida was removed, one embryo per well is recommended.

3.2.1 Whole Embryo Immunolabeling

1. Permeabilize the fixed embryos for 1 h at RT (*see* **Note 4**).

2. Carefully transfer embryos directly into blocking solution and incubate for 30 min at RT.

3. Incubate embryos with primary antibody against UBF diluted in blocking solution for 1 h at RT (*see* **Note 5**).

4. Remove the excess primary antibody by washing the embryos (washing solution) for 3 × 15 min at RT.

5. Incubate embryos with goat anti-mouse secondary antibody conjugated with Alexa Fluor 594/PBS for 1 h at RT (*see* **Note 6**).

6. Remove the excess secondary antibody by washing the embryos (washing solution) for 3 × 15 min at RT.

7. Incubate embryos with primary antibody against B23 diluted in blocking solution for 1 h at RT.

8. Remove the excess primary antibody by washing the embryos (washing solution) for 3 × 15 min at RT.

9. Incubate embryos with goat anti-mouse secondary antibody conjugated with Alexa Fluor 488 diluted in PBS for 1 h at RT.

10. Remove the excess secondary antibody by washing the embryos (washing solution) for 3 × 15 min at RT.

11. Transfer the embryos into 5–10 µl drop of fluorescent mounting media placed in the hole of the sheet reinforcements ring label glued on glass slide (*see* **Note 7**). Carefully place the cover slip over the drop by continuous observation of embryo under the dissecting microscope. If visualization of nucleus is required, embryos can be incubated for 2–3 min in Hoechst or DAPI solution prior to mounting.

12. Proper negative immunostaining control experiments should be performed: First stage controls (control of specificity and unspecific binding of the primary antibody), where the primary antibody is either preabsorbed by the corresponding peptide or where the primary antibody is substituted by an isotype antibody, and second stage controls, (control of the unspecific binding of the secondary antibody) where the primary antibody is omitted.

3.2.2 Embryo Processing for Transmission Electron Microscopy

1. Prepare 4 % agar solution in hot Milli-Q water in the Erlenmeyer flask and after dissolving the agar solution in microwave oven, place the flask into the thermic heater adjusted to

Fig. 5 Rotator containing glass containers used during embryo processing for transmission electron microscopy

50–60 °C. Cover the flask with aluminum foil and let it stand for 1 h in order to remove air bubbles (*see* **Note 8**).

2. Under continuous observation of manipulation under dissecting microscope, place a drop of agar over the fixed embryo positioned in the middle of a glass slide (*see* **Note 9**).

3. After solidification of agar, cut out the cubes containing the embryo (approximately $2 \times 2 \times 2$ mm) with scalpel. Embedding into agar allows for easy manipulation with the sample and avoids the use of microscopy in later steps, when toxic ingredients are used. The agar cubes can be stored in 0.1 M sodium-phosphate buffer at 4 °C or directly processed for TEM.

4. Wash the agar-embedded embryos 2×5 min in 0.1 M sodium phosphate buffer on rotator (4 rpm) (Fig. 5). Remove the fluid with disposable pipette to avoid the unwanted removal of agar cubes. We recommend using glass containers with flat bottom and cover with corks for all the steps prior to final embedding into epon, and applying 2–3 ml of solutions.

5. Post-fix the embryos for 1 h at RT under rotation (*see* **Note 10**). Cover the container with foil when fixation is carried out (*see* **Note 11**).

6. Wash the agar embedded embryos 2×5 min in distilled water.

7. Dehydrate the agar-embedded embryos in ascendant ethanol row under rotation: 10 min in 50 % ethanol, 10 min in 70 %

ethanol, 10 min in 96 % ethanol, 2×20 min in 99 % ethanol, and 20 min in ultrapure 99 % ethanol.

8. Remove the ultrapure ethanol and wash the samples 2×10 min in propylene oxide. Remove the fluid.

9. Add mixture of Propylene oxide–Epon (2:1) and infiltrate for 20 min on rotator. Remove the fluid.

10. Add mixture of Propylene oxide–Epon (1:1) and infiltrate for 20 min on rotator. Remove the fluid.

11. Add mixture of Propylene oxide–Epon (1:2) and infiltrate for 20 min on rotator. Remove the fluid.

12. Add pure Epon and infiltrate the embryos overnight on rotator. Do not cover the containers to allow for evaporation of propylene oxide.

13. Fill in fresh Epon in the holes in the mold (20 holes per mold) and transfer the agar cube with the embryo on top of the hole filled with Epon. The cube sinks to the bottom. Incubate for 24 h at 60 °C (*see* **Note 12**)

14. The Epon blocks are ready for serially semi-thin sectioning (2 μm) with a glass knife on an Ultramicrotome. Stain the semi-thin sections with toluidine blue and evaluate by bright-field light microscopy to assure the desired stages of embryonic development and to select sections for further processing.

15. Reembedding of the selected semi-thin sections: an area of approximately 2×2 mm containing the semi-thin section, is framed with nail varnish. Subsequently, a block of polymerized Epon, which has been trimmed at one end to an area of approximately 1×1 mm, is glued upright on the semi-thin section by means of a small drop of fresh Epon adhering the 1×1 mm area of the block to the semi-thin section. Following polymerization for an additional 1 day at 60 °C, the Epon block with the semi-thin section on the tip can be cracked off the slide by plunging into liquid nitrogen.

16. Ultrathin sections (40–60 nm) are cut using a diamond knife, resulting in grey-to-silver colored sections when observing the interference color. The sections are collected on 150 mesh copper grids coated with collodion.

17. The ultrathin sections are contrasted with uranyl acetate and lead citrate on the grids according to standard procedure.

3.3 Assessment of Nucleolar Reprogramming

3.3.1 Assessment Based on Localization of Nucleolar Proteins

Nucleolar proteins involved in rRNA synthesis (e.g., UBF) and processing (e.g., B23) display specific dynamics in their localization during early embryo development. In the first cell cycle(s), UBF and B23 are widely dispersed in cytoplasm as well as nucleoplasm (Fig. 4). Later, their localization becomes mainly nuclear, but remains dispersed. Around the time of genome activation, firstly processing and subsequently transcription related nucleolar proteins gain their specific localization. Processing proteins appear in a ring/shell like structures around the presumptive nucleoli. Proteins involved in rDNA transcription resemble small foci spread in nucleoli.

3.3.2 Assessment Based on Transmission Electron Microscopy of Nucleolar Ultrastructure

The ultrastructure of the nucleolar compartment is assessed by transmission electron microscopy at appropriate magnifications in order to visualize nucleolar sub-compartments as the FCs, the DFC, and the GC. Nucleolar ultrastructure is classified as being either inactive (nucleolar remnants or NPBs), in the process of nucleolar formation (internal or external nucleologenesis; for example of the former *see* Fig. 3), or active fibrillo-granular (presenting FCs, DFC, and GC, for examples *see* Figs. 1 and 3).

4   Notes

1. Serum may interact with fixative and decrease the quality of fixation.

2. The need for removal of zona pellucida (ZP) depends on the species and production method. For example in vitro produced embryos have different ZP composition which allows for immunolabeling after permeabilization. On the other hand, in vivo developed porcine embryos cannot be reliably labelled without ZP removal. The routine removal of ZP as used in different laboratories is sufficient.

3. Prepare 3 % glutaraldehyde from 25 % glutaraldehyde immediately before use to avoid the conversion of glutaraldehyde to succinic acid.

4. The concentration of Triton depends on presence/absence of ZP but may also vary depending on the species. For example embryos without ZP require only permeabilization with 0.1 % Triton for 30 min–1 h, while in vitro produced porcine embryos are permeabilized by 1–2 % Triton applied for 1–2 h.

5. The concentration of antibody may vary on the batch and should be adjusted.

6. The concentration of secondary antibody should be adjusted. The choice of fluorochrome depends on individual preferences. The type of the antibody should be adjusted according to the species examined.

7. Apply the ring labels with gloves, as fingerprints may interfere with acquisition of fluorescent pictures.

8. Bubbles may interact with later steps and markedly decrease the quality of the samples.

9. Avoid to transfer a large amount of buffer, since it may result in the embryo moving to the edge of the drop.

10. Most of the chemicals are toxic and all the steps after agar embedding should be performed in the fume-hood.

11. Collect the used fixative in a separate container and make it non-harmful with corn oil.

12. Manipulation of fragile Epon-impregnated agar cubes is difficult due to the Epon viscosity. Empty the glass container on filter pater and clean out the container with wooden skewer. Gently spread the fluid over the paper so that the agar cubes become visible. The skewer is also used for transfer of individual cubes into the Epon block forms.

References

1. Shaw PJ, Jordan EG (1995) The nucleolus. Annu Rev Cell Dev Biol 11:93–121

2. Scheer U, Hock R (1999) Structure and function of the nucleolus. Curr Opin Cell Biol 11: 385–390

3. Wachtler F, Stahl A (1993) The nucleolus: a structural and functional interpretation. Micron 24:473–505

4. Biggiogera M, Malatesta M, Abolhassani-Dadras S et al (2001) Revealing the unseen: the organizer region of the nucleolus. J Cell Sci 114:3199–3205

5. Koberna K, Malinsky J, Pliss A et al (2002) Ribosomal genes in focus: new transcripts label the dense fibrillar components and form clusters indicative of "Christmas trees" in situ. J Cell Biol 157:743–748

6. Flechon JE, Kopecny V (1998) The nature of the "nucleolus precursor body" in early preimplantation embryos: a review of fine-structure cytochemical, immunocytochemical and autoradiographic data related to nucleolar function. Zygote 6:183–191

7. Maddox-Hyttel P, Bjerregaard B, Laurincik J (2005) Meiosis and embryo technology: renaissance of the nucleolus. Reprod Fertil Dev 17:3–14

8. Hyttel P, Laurincik J, Rosenkranz C et al (2000) Nucleolar proteins and ultrastructure in pre-implantation porcine embryos developed in vivo. Biol Reprod 63:1848–1856

9. Laurincik J, Thomsen PD, Hay-Schmidt A et al (2000) Nucleolar proteins and nuclear ultrastructure in pre-implantation bovine embryos produced in vitro. Biol Reprod 62:1024–1032

10. Laurincik J, Schmoll F, Mahabir E et al (2003) Nucleolar proteins and ultrastructure in bovine in vivo developed, in vitro produced, and parthenogenetic cleavage-stage embryos. Mol Reprod Dev 65:73–85

11. Thompson EM (1996) Chromatin structure and gene expression in the preimplantation mammalian embryo. Reprod Nutr Dev 36: 619–635

12. Hay-Schmidt A, Viuff D, Greve T et al (2001) Transcriptional activity in in vivo developed early cleavage stage bovine embryos. Theriogenology 56:167–176

13. Ogushi S, Palmieri C, Fulka H et al (2008) The maternal nucleolus is essential for early embryonic development in mammals. Science 319:613–616

14. Gavrilova EV, Kuznetsova IS, Enukashvili NI et al (2009) Localization of satellite DNA and associated protein in respect to nucleolar precursor bodies in one- and two-cell mouse embryos. Tsitologiia 51:455–464

15. Fulka H, Fulka J Jr (2010) Nucleolar transplantation in oocytes and zygotes: challenges for further research. Mol Hum Reprod 16:63–67

16. Martin C, Beaujean N, Brochard V et al (2006) Genome restructuring in mouse embryos during reprogramming and early development. Dev Biol 292:317–332

17. Martin C, Brochard V, Migne C et al (2006) Architectural reorganization of the nuclei upon transfer into oocytes accompanies genome reprogramming. Mol Reprod Dev 73: 1102–1111

18. Ahmed K, Dehghani H, Rugg-Gunn P et al (2010) Global chromatin architecture reflects pluripotency and lineage commitment in the early mouse embryo. PLoS One 5:e10531

19. Pichugin A, Le BD, Adenot P et al (2010) Dynamics of constitutive heterochromation: two contrasted kinetics of genome restructuring in early cloned bovine embryos. Reproduction 139:129–137

20. Hyttel P, Laurincik J, Zakhartchenko V et al (2001) Nucleolar protein allocation and

ultrastructure in bovine embryos produced by nuclear transfer from embryonic cells. Cloning 3:69–81

21. Svarcova O, Strejcek F, Petrovicova I et al (2008) The role of RNA polymerase I transcription and embryonic genome activation in nucleolar development in bovine preimplantation embryos. Mol Reprod Dev 75:1095–1103

22. Deshmukh RS, Østrup O, Strejcek F et al (2012) Early aberrations in chromatin dynamics in embryos produced under *in vitro* conditions. Cell Reprogram 14:225–234

23. Laurincik J, Bjerregaard B, Strejcek F et al (2004) Nucleolar ultrastructure and protein allocation in in vitro produced porcine embryos. Mol Reprod Dev 86:327–334

24. Takahashi K, Yamanaka S (2006) Induction of pluripotent stem cells from mouse embryonic and adult fibroblasts cultures by defined factors. Cell 126:663–676

25. Vierbuchen T, Ostermeier A, Pang ZP et al (2010) Direct conversion of fibroblasts to functional neurons by defined factors. Nature 463:1035–1041

26. Østrup O, Hyttel P, Klærke DA et al (2011) Remodeling of ribosomal genes in somatic cells by *Xenopus* egg extract. Biochem Biophys Res Commun 412:487–493

27. Murayama K, Ohmori A, Fujimura H et al (2008) Epigenetic control of rDNA loci in response to intracellular energy status. Cell 133:627–639

28. Preuss S, Pikaard CC (2007) rRNA gene silencing and nucleolar dominance: insights into a chromosome-scale epigenetic on/off switch. Biochim Biophys Acta 1769: 383–392

29. Santoro R, Grummt I (2005) Epigenetic mechanism of rRNA gene silencing: temporal order of NoRC-mediated histone modification chromatin remodeling and DNA methylation. Mol Cell Biol 25:2539–2546

30. Glauert AM (1975) Fixation, dehydration and embedding of biological specimens. American Elsevier, New York

31. Reynolds ES (1963) The use of lead citrate at high pH as an electron-opaque stain in electron microscopy. J Cell Biol 17:208–212

Chapter 13

Assessment of Cell Lineages and Cell Death in Blastocysts by Immunostaining

Sabine Chauveau and Claire Chazaud

Abstract

During the last decade it has been shown that most mammalian blastocysts consisted of three cell lineages. Immunofluorescence with multiple antibodies enables to identify each cell type allowing an easy detection of eventual defects. It is complementary to RT-PCR experiments as this technique allows to look at cell position and to analyze and count the proportions between the different cell types. Thus after any kind of embryo manipulation such as nuclear transfer (NT), the analysis of the three cell lineages by immunofluorescence will provide criteria for good or poor development.

Key words Immunofluorescence, Blastocyst, Cell lineages, Cell death detection, Confocal microscopy

1   Introduction

Early in development, mammalian embryos establish extraembryonic tissues to prepare for nutritional needs while developing pluripotent cells for the formation of the embryo proper. The first cell lineage differentiation separates inner cell mass (ICM) cells from trophectoderm (TE) cells. Due to their position and polarization, the outer cells of the late morula acquire a TE identity, visualized by the expression of Cdx2, Gata3 and, later, Eomesodermin [1]. ICM cells on the other hand express Oct4. In mice, while Cdx2, Gata3, and Oct4 are initially coexpressed from around the 8-cell stage, their expression is clearly restricted to the TE or the ICM respectively at the blastocyst stage. Thus, the presence of ambiguous cells could be a sign of developmental delay or defect. After TE specification, ICM cells differentiate into Epiblast (Epi) and Primitive Endoderm (PrE) lineages. By the late blastocyst stage in the mouse, Epi- and PrE-precursor cells are intermingled within the ICM in a "salt and pepper" pattern. Up to now, cell fate was not clearly associated to cell position and actual models present a stochastic activation of cell specification. *Nanog* and *Gata6* were the first genes to show an exclusive and reciprocal expression pattern

Nathalie Beaujean et al. (eds.), *Nuclear Reprogramming: Methods and Protocols*, Methods in Molecular Biology, vol. 1222, DOI 10.1007/978-1-4939-1594-1_13, © Springer Science+Business Media New York 2015

by E3.5 in the mouse [2] while being coexpressed at earlier stages [3]. Between the late morula stage (E3.0) and the late blastocyst stage (E3.75), individual ICM cells select their identity by downregulating either *Nanog* or *Gata6* expression [4]. However cells are not synchronized, some ICM cells being specified later than others, yielding to difficult analyses. In spite of that, cells are generally specified by E3.75. Differences in the populations proportions has been observed between mouse strains but are generally close to 50 % for Epi (50–40 % of ICM cells) and PrE (50–60 %) [5]. Thus, when analyzing other mammalian species or strains, it is advised to examine cell numbers of on freshly collected blastocysts and compare to NT embryos. The proportions between these two cell populations might be important for the correct development of the embryo. Indeed, these cell lineages require each other for their further development and differentiation. For example, PrE needs Fgf4 secreted from the Epi cells to be able to express Gata4 and Sox17 that are later markers of PrE differentiation [5, 6]. Once ICM cells are specified, the two populations sort to give rise to two distinct tissues by E4.0, the PrE epithelium separating the Epi from the blastocyst cavity. The correct segregation of ICM cells is another criterion of correct development.

Thus it is important that cells select their identity properly and in correct proportions of the three lineages. Analyses by RT-PCR have been done on whole NT embryos as the embryo is small ($\approx$80 μm wide) at that stage [7, 8], but, as the different cell types are intermingled, it is difficult to assess the quality. Single cell RT-PCR techniques are available to study blastocysts [4] and are suitable to quantify each marker; however, in situ analyses remain essential to observe cell repartition and numbers.

The following protocol is suitable to many (but not all) antibodies, although to identify the three cell types, we advise to use Cdx2 (TE), Nanog (Epi), and Gata6 (PrE) or Sox17 (PrE) antibodies. They are all transcription factors that label the nucleus, making it easy for detection and cell count. As the three cell lineages must be detected on the same samples, antibodies produced in different animals are to be used to avoid cross-reactions. A TUNEL reaction can be added to detect cell death. Note that as TUNEL detects fragmented DNA, nucleus integrity is lost in the dying cells and it is thus impossible to determine the identity of the cell with the above markers.

Here we describe the protocol routinely carried out on mouse preimplantation embryos. Briefly embryos are collected and fixed with paraformaldehyde. They are then incubated with the primary antibodies, all at the same time and afterwards with secondary antibodies. A TUNEL reaction can be added as a last step. Observation is then performed on a confocal microscope to be able to distinguish the different nuclei within the volume of the embryo. Finally, after image acquisition, different software (manual or automated) can be used to count cell types.

For other mammals, this protocol has to be adapted (fixation, permeabilization …) and the antibodies that work well in the mouse need to be individually tested in other species.

2 Materials

1. Pulled Pasteur pipets and mouth tubing (*see* **Note 1**).
2. 4-well plates.
3. Nutator mixer.
4. 1× Phosphate Buffer Saline (PBS).
5. PFA: freshly dissolved 4 % paraformaldehyde in PBS (can be kept at +4 °C for 1 week).
6. PBT: 0.1 % Tween 20 in PBS (*see* **Note 2**).
7. Fetal Bovine Serum (FBS).
8. Primary antibodies directed against Nanog, Cdx2, Gata6, and Sox 17 (*see* **Note 3**).
9. Secondary antibodies directed against the different animals of the primary antibodies. Make sure that each antibody is coupled to a different fluorophore that can be distinguished by the confocal microscope that will be used. Generally, Alexa 488, Cy3, Cy5 suit most microscopes (*see* **Note 4**).
10. Dapi or Hoechst DNA stains.
11. TUNEL reaction kit that allows cell death detection (*see* **Note 5**).
12. Coverslips for confocal microscopy.
13. Isolators: Press-*to-seal*™ (P24743, Invitrogen) to avoid squashing the embryos during image acquisition.
14. Inverted confocal microscope.
15. Image analysis software (*see* **Note 6**).

3 Methods

Unless otherwise indicated, experiments are performed at room temperature (RT) in 500 μl solutions in 4-well plates. Except for mouth pipet transfers, embryos are kept in the well: the old solution is removed and replaced by the next one (gentle use of micropipets). During RT incubations and washes we use a nutator mixer that gradually places the embryos at the center of the well, enabling an easier pipeting and avoiding loss of samples.

3.1 Fixation

The protocol can be performed on freshly collected blastocysts or NT ones.

1. Briefly rinse the embryos in a 50 μl drop of PBS by mouth pipet transfer.

2. Transfer embryos into PFA. Fix 15 min at RT or overnight (O/N) at +4 °C (*see* **Note 7**).

3. Transfer in PBT. Wash three times 5 min in PBT.

3.2 Immunolabeling

4. Block antibody-aspecific sites with 10 % FBS in PBT for 30 min.

5. Incubate with primary antibodies O/N at +4 °C (*see* **Note 8**).

6. Wash with PBT for 5, 10, 15, 15 min.

7. Incubate for 30 min at RT with secondary antibodies and DNA dye, diluted in 10 % FBS in PBT at concentrations advised by manufacturers.

8. Wash twice 5 min in PBT.

3.3 TUNEL

For an additional TUNEL reaction using the In Situ Cell Death Kit (*see* **Note 9**):

1. Dilute the Enzyme Solution in the Label Solution at 1/10 (v/v). Mix well.

2. Transfer embryos in a 100 μl drop of this labeling solution. Incubate 1 h at 37 °C in a humidified chamber, in the dark.

3. Wash twice 5 min in PBT.

3.4 Mounting, Imaging, and Counting

1. Cut out and stick insulator wells onto a coverslip. Remove the protection film. Fill microwells with 52 μl of PBT.

2. Transfer embryos into a well. Cover with another coverslip, avoiding bubbles.

3. Confocal image acquisition can then be performed (*see* **Note 10**) using ×40 objectives (Fig. 1). Make z-sections every 2 μm.

4. Analyze/count each cell for its identity (single or colocalization of different labellings) (*see* **Note 11**).

4   Notes

1. To prepare mouth pipets, stretch a glass capillary or a Pasteur pipet in the flame. Mouth tubing can be hand-made but are also commercially available. For more details refer to Chapter 11 and to Nagi and collaborators [9].

2. Tween can be replaced by Triton X100 at each PBT step.

3. We recommend using the following primary antibodies that provide good signals (Fig. 1): rabbit Nanog (ab80892, Abcam, UK); mouse Cdx2 (ab86949, Abcam, UK); goat Gata6 (AF1700, R&D Systems, MN, USA); goat Sox17 (AF1924, R&D Systems, MN, USA).

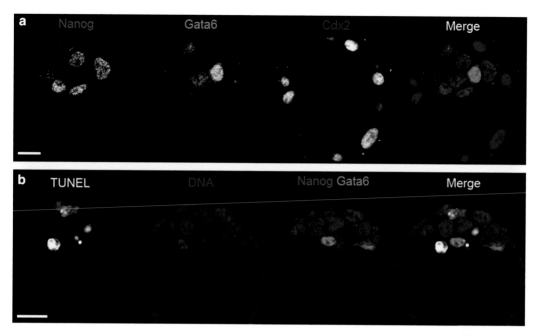

Fig. 1 Labeling of E3.75 mouse embryos. (**a**) Immunostaining with Nanog (Cy3), Gata6 (Alexa-488), and Cdx2 (Cy5). (**b**) TUNEL, DNA, and immunofluorescence stainings. On this z-section four cells are apoptotic. Scale bars 20 μm

4. For the primary antibodies cited in **Note 3** you will need the following secondary antibodies: anti-mouse, -rabbit, -goat made in donkey. We usually work with antibodies from Jackson ImmunoResearch Laboratories (PA, USA) at 1/300 dilution.

5. We usually use the In Situ Cell Death Kit from Roche Applied Science.

6. ImageJ free software for example can be downloaded at http://rsbweb.nih.gov/ij/. IMARIS from Bitplane is another, more sophisticated, software dedicated to image restoration, visualization, and analysis.

7. We have not seen any difference between 15 min RT and O/N at +4 °C for the described antibodies Embryos can be stored a couple of days in PFA but it is preferable not to.

8. Dilute Nanog and Gata6 or Sox17 antibodies (*see* **Note 4**) at 1/300 into the Cdx2 antibody solution which is provided at 1×. This Sox17 antibody makes a brighter signal than Gata6's. However, Sox17 is only suitable for late blastocyst stages as it is a later PrE marker than Gata6 in the mouse.

9. TUNEL reaction will occupy one channel during confocal imaging. Depending on your microscope and antibodies/fluorophores availability, it will be possible to carry out four-channel acquisition (five with DNA stainings). Otherwise, only two antibodies will be chosen for the experiment.

10. As embryos are mounted in PBT, image acquisition has to be done within 24 h. Alternatively mounting media can be used.

11. This step can be manually carried out using the "multi-point selection" tool of ImageJ section by section or using more sophisticated tools, for example from IMARIS (Bitplane).

Acknowledgement

This work was supported by the ANR "EpiNodal."

References

1. Stephenson RO, Rossant J, Tam PP (2012) Intercellular interactions, position, and polarity in establishing blastocyst cell lineages and embryonic axes. Cold Spring Harb Perspect Biol 4. doi:10.1101/cshperspect.a008235

2. Chazaud C, Yamanaka Y, Pawson T et al (2006) Early lineage segregation between epiblast and primitive endoderm in mouse blastocysts through the Grb2-MAPK pathway. Dev Cell 10:615–624

3. Plusa B, Piliszek A, Frankenberg S et al (2008) Distinct sequential cell behaviours direct primitive endoderm formation in the mouse blastocyst. Development 135:3081–3091

4. Guo G, Huss M, Tong GQ et al (2010) Resolution of cell fate decisions revealed by single-cell gene expression analysis from zygote to blastocyst. Dev Cell 18:675–685

5. Frankenberg S, Gerbe F, Bessonnard S et al (2011) Primitive endoderm differentiates via a three-step mechanism involving Nanog and RTK signaling. Dev Cell 21:1005–1013

6. Kang M, Piliszek A, Artus J, Hadjantonakis AK (2013) FGF4 is required for lineage restriction and salt-and-pepper distribution of primitive endoderm factors but not their initial expression in the mouse. Development 140:267–279

7. Sebastiano V, Gentile L, Garagna S et al (2005) Cloned pre-implantation mouse embryos show correct timing but altered levels of gene expression. Mol Reprod Dev 70:146–154

8. Jouneau A, Zhou Q, Camus A et al (2006) Developmental abnormalities of NT mouse embryos appear early after implantation. Development 133:1597–1607

9. Nagy A, Gertsenstein M, Vintersten K et al (2003) Manipulating the mouse embryo: a laboratory manual. Cold Spring Harbor Press, Cold Spring Harbor, NY

Chapter 14

Gene Expression Analysis in Early Embryos Through Reverse Transcription Quantitative PCR (RT-qPCR)

Nathalie Peynot, Véronique Duranthon, and Daulat Raheem Khan

Abstract

Real-time, reverse transcription quantitative PCR (RT-qPCR) is a highly sensitive and reproducible technology for the analysis of gene expression patterns. Its ability to detect minute quantities of nucleic acid from multifarious sources makes it an ideal technique for embryonic transcript quantification. However, complex cellular diversity and active transcriptome dynamics in early embryos necessitate particular caution to avoid erroneous results. This chapter is intended to outline basic methodology to design and execute RT-qPCR experiments in pre-implantation embryos.

Key words Embryo, Gene expression, Quantitative-PCR, RT-qPCR, Data analysis, Reference genes

1 Introduction

Over the past 20 years the application of RT-qPCR has greatly increased. The gene expression profiles in tissues or cells could be robustly analyzed in various fields of biology using this highly sensitive technique. However, due to the specificities of early mammalian embryos, technical adjustments are required to avoid erroneous results produced by this technology. In fact, the embryo is a highly dynamic model in terms of total RNA content, gene expression, cell number and the state of cell differentiation. Moreover, it contains very small amounts of biological material (tens of pictograms messenger RNA per embryo). The degree of complexity further increases with an inter-embryo variability, which could, at least partly, be related to the genetic variability. Therefore, these properties have to be taken into account for the careful establishment of gene expression profiles in early embryos. The aim of this chapter is to describe a procedure for robust comparison of gene expression between several pools of preimplantation mammalian embryos. Here, we shall describe a protocol for relative quantification of gene expression using the delta-delta Cq method which is associated with standard curve determination of PCR efficiencies.

Nathalie Beaujean et al. (eds.), *Nuclear Reprogramming: Methods and Protocols*, Methods in Molecular Biology, vol. 1222, DOI 10.1007/978-1-4939-1594-1_14, © Springer Science+Business Media New York 2015

In this method, results are expressed as the ratio of expression of the gene of interest between the biological sample and a calibrator sample (usually a reference stage of embryonic development or a control) divided by similar ratio for the reference genes.

1.1 The Procedure Principle

Following RNA extraction, the transcript quantification in early embryos through RT-qPCR, is achieved by a two-step strategy which involves *first*, reverse transcription of embryonic RNA into cDNA and *second*, exponential amplification and simultaneous measurement of accumulating cDNA using polymerase chain reaction (PCR). In fact, the accumulation of PCR product is measured by the increase in a fluorescent signal to each sample after each amplification cycle. This fluorescence is produced either by a nonspecific fluorophore which intercalates in the double stranded DNA molecule (e.g., SYBR Green) or by hybridization and/or hydrolysis of sequence-specific fluorescent probes during the qPCR reaction. Later on, after an adequate normalization by reference genes, the differences in the fluorescence data across samples are used to estimate the inherent variations of transcript abundance amongst experimental conditions [1]. The value of fluorescence signal generated by qPCR instrument is the number of PCR cycles required for fluorescence signal to reach the threshold of fluorescence detection. This is usually referred to as Quantification Cycle (Cq) or Threshold Cycle (Ct). In PCR amplification, the fluorescence from a more concentrated standard solution or sample with higher gene expression will reach the threshold at an earlier cycle (that is with a lower Cq) compared to a less concentrated solution or sample with lower gene expression level (otherwise a higher Cq).

In order to achieve optimum results, similar conditions of reverse transcription are necessary across all the samples to be compared. Indeed, the efficiency of the reverse transcription reaction depends on the priming strategy and total RNA concentration in the sample [2]. In early embryos, precise RNA quantification after extraction is difficult due to small RNA quantities. In order to circumvent this problem, we start with the same number of embryos in all samples of an experiment, and we add the same amount of carrier RNA in all samples at the very beginning of the extraction procedure. The quantity of embryonic RNA is negligibly small compared to the carrier RNA added to a sample, however, reverse transcription efficiency of this minute embryonic RNA remains unaffected by these carrier molecules. The addition of this carrier RNA to samples is required for an easy and reliable estimation of the extraction yield. The extrapolation of this extraction yield to the initial embryo number enables to calculate an "equivalent embryo" number available after extraction for each sample. Subsequently, the reverse transcriptase reactions will be performed on the same number of "equivalent embryo" for all the samples of the same experiment.

qPCR normalization also requires the use of reference genes whose expression is stable amongst the samples to be compared. Such genes are very difficult to find during preimplantation development due to the decrease in maternal transcripts and increase in embryonic ones. In order to address this problem, an exogenous reporter RNA is added in a constant amount per embryo in the samples at the beginning of RNA extraction procedure. This exogenous reporter RNA will be used only if, at the end of the whole procedure, the expected endogenous reference genes do not satisfy the criteria for a robust normalization.

Based on the above perspective, we describe here various steps involved in application of RT-qPCR for gene expression analysis in early embryos.

2 Materials

2.1 Equipment

- Binocular loop.
- Water bath and heating block.
- Pipettes and low retention filter tips of capacities 2, 10, 20, 200, and 1,000 μL.
- RNase-free microcentrifuge tubes, 600, 1,500 μL (low binding for nucleic acid).
- Optical 96 well reaction plate and Optical adhesive Film or Fast 8 tube strip and Optical 8-cap strip.
- Benchtop centrifuge.
- UV spectrophotometer.
- Real-time qPCR thermal cycler compatible with the choice of fluorescent system.

2.2 RNA Isolation

- Carrier RNA (e.g., 16S and 23S Ribosomal RNA).
- RNA of exogenous reporter gene (e.g., luciferase RNA).
- Column based RNA isolation kit.
- DNase treatment kit for column extraction (e.g., RNase-free DNase kit from Qiagen).
- DEPC-treated water.

2.3 Reverse Transcription

- Random oligonucleotide hexamer primers or Oligo d(T) Primers.
- Deoxyribonucleotides (dNTPs) in water (10 mM).
- A reverse transcription enzyme operating at a high temperature and having a high efficiency for low amounts of material (e.g., Superscript III from Invitrogen).
- RNase inhibitor 40 U/μL.

2.4 Standard Solutions Preparation

– Quantitect® Whole Transcriptome kit from Qiagen.

2.5 Primer or Probe Design and Quantitative PCR Performance

– cDNA sequence of the genes to be analyzed. Furthermore, location of exon junctions and any alternate splice forms should also be considered.

– Primers and probes assay may be designed by using design programs provided by the equipment supplier or using free software available on the Web.

– Preferentially use master mix compatible with SYBR Green or fluorescent probe system.

3 Methods

The standard method to perform RT-qPCR analysis (charted as Fig. 1) include *sample preparation* (sample collection, RNA isolation and reverse transcription), *assay design* (choice of fluorescent system, primer/probe design and preparation of standard solutions), *assay performance* (performance of qPCR and assay validation), and *data analysis* (normalization against reference genes and statistical analysis).

3.1 Sample Preparation

3.1.1 Embryo Collection

1. In order to perform gene expression analyses, embryos should be collected in RNase and DNase-free environment, dry-frozen and stored until use. First of all, wash the embryos twice in PBS solution (warmed to 37 °C) to remove any traces of

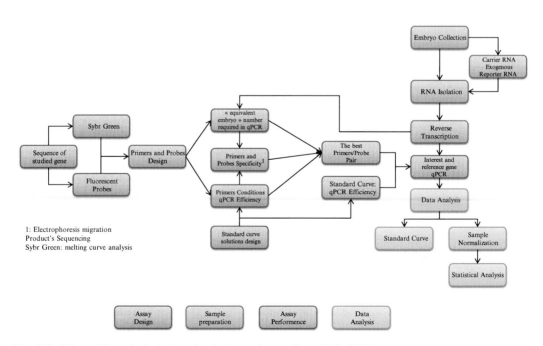

Fig. 1 Workflow of the principal steps for design and execution of RT-qPCR in early embryos

culture medium. A 4-well plate may be used for this purpose containing 500 μL of PBS in each well.

2. Transfer the washed embryos to correctly labelled, RNase-free microcentrifuge tubes (low binding for nucleic acid tubes) with minimum possible solution of PBS (*see* **Note 1**). Under binocular, remove the PBS solution with a small pipette.

3. Freeze the embryos immediately in liquid nitrogen or dry ice and store at −80 °C.

3.1.2   RNA Isolation

RNA isolation is performed through several methods; however, solid phase RNA isolation, as adsorption chromatography, is recommended for embryonic gene expression analyses (*see* **Note 2**). In adsorption chromatography, nucleic acids are selectively bound by adsorption on silica or glass in the presence of certain salts (including chaotropic salts), while other biological molecules do not bind. Buffer or low salts water can then elute the nucleic acids and thereby produce a sample to be used directly in downstream applications. We choose to use Arcturus PicoPure RNA Isolation Kit (Applied Biosystems) but other extraction kits adapted for small samples can be used.

The protocol described here shows the best way to integrate Carrier RNA (to evaluate the extraction yield) and exogenous reporter RNA (to normalize the qPCR results) during extraction protocol.

During the extraction procedure, we recommend including a treatment step with DNAse to remove all traces of genomic DNA.

1. Get the samples from −80 °C freezer just before starting the extraction and keep them on dry ice until **step 2**.

2. Add carrier RNA (2.5 μg/sample) and RNA of exogenous reference gene (e.g., Luciferase 1 pg/embryo) into each sample just before adding the extraction buffer. The quantity of carrier RNA remains constant for all samples; the quantity of exogenous reporter gene RNA depends upon the number of embryos in each tube (*see* **Note 3**).

3. Add 50–100 μL of Extraction Buffer (XB) directly to the sample and mix gently by pipetting up and down and incubate at 42 °C for 30 min in water bath or thermal cycler.

4. At the end of the above step, pre-condition the RNA Purification Column by pipetting 250 μL of Conditioning Buffer (CB) onto the purification column filter membrane, incubate for 5 min at room temperature, and centrifuge the purification column at $16,000 \times g$ for 1 min.

5. Following the 42 °C incubation step, quick spin the tube to collect all the liquid at the bottom of the vial.

6. Add an equal volume of 70 % ethanol to the embryonic cell extract (50–100 μL of XB + volume of carrier RNA + volume of

exogenous reporter RNA). Mix by pipetting up and down. Do not centrifuge.

7. Add the cell extract and ethanol mixture onto the preconditioned purification column, centrifuge for 2 min at $100 \times g$ and immediately followed by $16,000 \times g$ for 30 s.

8. Add 100 μL of Wash Buffer1 (W1) onto the purification column and centrifuge for 1 min at $8,000 \times g$.

9. Perform DNase treatment to remove any DNA contamination. Pipette the 40 μL DNase I diluted in RDD buffer (e.g., RNase-Free DNase set, Qiagen®) (see **Note 4**) and incubate at 20–30 °C for 15 min. Add 40 μL Wash Buffer1 (W1) onto the purification column membrane and centrifuge at $8,000 \times g$ for 15 s.

10. Wash the RNA sample twice with 100 μL Wash Buffer2 (W2). After first wash centrifuge the tubes for 1 min at $8,000 \times g$ and after second wash centrifuge for 2 min at $16,000 \times g$. Check the purification column for any residual wash buffer. If wash buffer remains re-centrifuge at $16,000 \times g$ for 1 min.

11. Transfer the columns on collection tubes and add 13 μL of Elution Buffer (EB) directly onto the membrane of the purification column. Incubate the purification column for 1 min at room temperature and centrifuge initially for 1 min at $1,000 \times g$ and then for 1 min at $16,000 \times g$.

12. Measure the volume of the sample.

13. Use 1.5 μL of sample to determine the concentration of RNA in order to estimate the extraction yield (Table 1). This yield is used to estimate the number of "equivalent embryos" in a sample (see **Note 5**).

14. Store the sample at –80 °C until further step.

3.1.3 Reverse Transcription

It is well understood that a PCR reaction requires "DNA" as template. Therefore, to quantify transcripts by qPCR, a reverse transcription (RT) reaction is performed to convert RNA samples

Table 1
Estimation of RNA isolation efficiency

Number of embryos	Quantity exogenous reporter RNA (pg)	Quantity carrier RNA (ng)	Volume eluate (μL)	Remaining volume after DO (μL)	Concentration (ng/μL)	Extraction yield (%)	Equivalent embryo in remaining volume
q	q	Q	V	v	C	C × V × 100/Q	C × q × v/Q
16	16	2,500	13	11	174	90.48	12.24
18	18	2,500	13	11	175	91	13.86

into complementary DNA (cDNA) before executing qPCR. For all samples, an equal number of equivalent embryos should preferably be used to avoid any technical variation and the reverse transcription reaction must be initiated with the same type of primers (*see* **Note 6**).

We show here the protocol for Superscript III Reverse Transcriptase (Invitrogen) with random primers annealing but other enzymes having the same properties as this product can be used.

1. Aliquot the RNA samples in RNase-free PCR tubes. It is preferred to use same number of equivalent embryos across different samples to avoid technical differences. Adjust the volume to 11 μL with DEPC-treated water. Add a negative control without RNA (*see* **Note 7**).

2. Add 1 μL (250 ng/μL) random hexamer and 1 μL dNTPs (10 mM), to each sample, mix and spin centrifuge to collect the reaction mixture at the bottom of the tube. It is recommended to prepare a master mix in a separate tube and distribute 2 μL to each sample in order to minimize technical variations.

3. Incubate the samples at 65 °C for 5 min in a heating block or thermocycler and then immediately cool on ice for 2–3 min. Quick spin the tubes to collect all the liquid at the bottom of the vial.

4. Add to each sample 4 μL 5× Reaction Buffer, 1 μL DTT (0.1 M), 1 μL RNasin (40 U/μL), and 1 μL SuperScript III reverse transcriptase (200 U). It is preferred to prepare a master mix in a separate tube and distribute 7 μL to each sample in order to minimize technical variation.

5. Mix tubes by gentle flicking and spin down the reaction mix at the bottom.

6. Place samples at 25 °C for 5 min. followed by an incubation of 50 °C for 1 h. in a thermal cycler or water bath.

7. Heat samples at 70 °C for 15 min and subsequently cool on ice for 2–3 min. Quick spin the tubes to collect all the liquid at the bottom of the vial.

8. Adjust reaction volume with DEPC-treated water (*see* **Note 8**) and store at –20 °C.

3.2   Assay Design

3.2.1   Choice of Florescent System

It is essential to design the RT-qPCR assay with extreme care to obtain reliable and interpretable results. In this regard, the first step involves the choice of an appropriate fluorescent system to be used. In fact, there are multiple fluorescent systems available, however, these may be divided into two main groups (a) nonspecific intercalating fluorophores like SYBR Green (*see* **Note 9**) and (b) sequence-specific fluorescent probes (*see* **Note 10**). Indeed, both these systems generate reliable results if proper controls are run along with. SYBR Green system is much less expensive whereas fluorescent probes are more specific and thus avoid chances of false positive results.

Here we detail the qPCR using SYBR Green because its procedure's development is easy and low cost.

3.2.2 Primer Design for Intercalating Dyes

Here we list the main characteristics to take into account for primer design. Each user can select software dedicated to primer design among all those available on the Web or sold by suppliers (*see* **Note 11**). It is preferred to design the primers on two different exons, where possible, to avoid genomic DNA amplification. Otherwise, the absence of fluorescent signal due to contaminant genomic DNA has to be controlled by the addition of negative RT controls (samples without reverse transcriptase enzyme).

In all cases, the design of primers depends on how the reverse transcriptase is initiated. In fact, if an oligo d(T) is used to initiate the reverse transcriptase reaction, it is better to design the primers and probes nearer to the RNA polyA tail. However, the use of random hexamer primers makes it possible to design the primers along all the RNA sequence.

3.2.3 Primer Design for SYBR Green Assay

1. Choose the product size range: 100–300 bp.
2. Primer length to be 18–20 bp.
3. Primer T_m around 60 °C with a range between 58 and 62 °C.
4. Select a pair of primers which has GC% not more than 50 % and minimum of secondary structure between the primers (hairpin loops, self dimers, or cross dimers between primers).
5. Blast the primers against the sequence of the species to be tested in NCBI Blast tool to ensure specificity.

3.3 Standard Solutions

During the assay design phase of qPCR conditions, the standard solutions are used to define the optimal primers conditions for a maximum PCR efficiency. Later, these are included in each assay to define the efficiency of the qPCR, which is required for data analysis. It is not necessary to determine the exact number of copies in the solution because results will be expressed as a ratio or as relative quantification of gene expression. Therefore, the relative dilutions of standard solutions should be known. In most of the qPCR software, by the use of standard solution relative dilutions, standard curve will be calculated automatically.

Alternatively, for each gene, a standard curve from the data of standard solutions may be generated manually by exporting the fluorescence data in the form of Cq values (also called Ct value) to a spread sheet.

Table 2
Analysis of standard curve of gene

Arbitrary units	Log	Mean Cq
1	0	NA
10	1	34.67
100	2	30.66
1,000	3	27.25
10,000	4	24.14
100,000	5	20.23
1,000,000	6	16.43
Slope	−3.5886	
Efficiency	90 %	

Standard Curve Analysis

$y = -3.5886x + 38.123$
$R^2 = 0.9988$

1. Manual calculation of the PCR efficiency E
   - Plot the Cq values against the log of the corresponding arbitrary scale representative for the serial dilution of the standard solution. Draw the regression line (standard curve) and calculate its slope. Note that the correlation coefficient should be close to 1.
   - Determine the PCR efficiency using the slope of the standard curve. A slope of −3.32 demonstrates 100 % efficiency; the experimental efficiency is calculated as follows (Table 2).

$$E\left(\%\right) = \left[\left(10^{(-1/\text{slope})}\right) - 1\right] \times 100$$

2. Preparation of standard solutions using pre-amplified RNA
   In order to determine relative quantification, any RNA or cDNA stock solution containing the appropriate target can be used to prepare standards. However, standard solutions should have a molecular complexity close to that of samples in order to assume a similar PCR efficiency. Embryos contain tiny amounts of RNA

which makes it difficult to create a standard curve without using large number of embryos. This can be overcome using pre-amplification of RNA from several embryonic stages to obtain standard solutions containing all embryonic transcripts from the oocyte stage to the blastocyst stage. We use the QuantiTect® Whole Transcriptome Kit by Qiagen for this purpose. This kit allows linear amplification and synthesis of double stranded cDNA through RNA reverse transcription using both oligo (dT) and random primers. Briefly the procedure is following.

– Extract Total RNA from large batches ($n = 40–50$) of embryos at different developmental stages as previously described, with exogenous reporter RNA but without carrier RNA—Determine the RNA quality and quantity by microfluidics technology (we use Bioanalyser).

– Submit a fixed quantity of each RNA sample to QuantiTect® Whole Transcriptome Kit reaction according to the manufacturer's instructions.

– Pool the different samples (equal volumes for each sample).

– Dilute this solution in water to obtain the standard solutions.

3.4 Optimization of PCR Conditions and Quality of the Design

In order to generate reliable results, validation of qPCR assay should be performed before proceeding to the actual samples.

1. *Primers conditions and qPCR efficiency*
   Using standard solutions, determine for each pair of primers the best experimental conditions. This includes determination of the concentration and annealing temperature for each pair of primers which results in the highest qPCR efficiency, the greatest sensitivity (determined by the lowest Cq for each standard solution) and the absence of dimers of primers.

2. *Specificity of the amplification*
   The analysis of melting or dissociation curve is necessary for nonspecific intercalating dyes assay. In fact, there are chances of secondary PCR products like primer-dimers which may affect transcript quantification. In case of primer specificity a single amplicon will be produced and hence, this will create a single dissociation peak. Otherwise, multiple peaks will be created. Most of the qPCR software produce a melting curve which may be consulted at the end of reaction. It is also necessary to control the specificity of the amplified product. For this, the size of the amplicon is checked by electrophoresis. The PCR product is purified and sequenced to confirm its identity (*see* **Note 12**).

3. *"Equivalent embryo" number required in qPCR*
   The number of "equivalent embryos" should remain constant across all samples to be compared for any particular gene otherwise efficiency of RT-PCR may change [2]. Before passing

the samples, the minimum number of "equivalent embryos" necessary for each embryonic stage and for each qPCR gene must be determined. The Cq values obtained by an optimum number of equivalent embryos should be in the linear part of the standard curve and preferably in the middle.

4. *Selection of Reference Genes in Embryonic Sample*
   Selection of an appropriate set of reference genes is the next most critical step which may affect the validity and reliability of results in qPCR experiments. Theoretically, levels of expression for the genes of interest are normalized to a set of three to five reference genes to correct the technical variations in sample preparation and to make comparisons across different conditions [3, 4]. Ideally, a reference gene should have a similar expression level across all conditions in an experiment. In practice, software such as geNorm calculate a parameter ("*M*" value) related to the stability of a potential reference gene and determine how many genes are required for reliable normalization [3]. Since embryo is an extremely dynamic entity in terms of gene expression patterns, it may be difficult to find such a set of endogenous reference genes. Therefore, an exogenous reporter transcript is added to the sample which is used to normalize the results of the genes of interest if an appropriate set of endogenous reference genes is not available (*see* **Note 13**). Obviously, this reporter transcript should not be expressed by the embryos [5, 6].

3.5 Assay Performance

3.5.1 Quantitative PCR

It is recommended to perform qPCR in triplicate reactions for each biological sample along with biological replicates of the experiments. QPCR is usually performed in 25 μL reaction volumes which include master mix (reaction buffer, dNTPs, $MgCl_2$, Taq polymerase, and primers) and RT product (*see* **Note 14**).
For each gene:

1. Prepare a master mix for each assay. Include an extra volume of at least one reaction to account for losses during manipulation (Table 3).

2. Distribute 15 μL of master mix to the relevant wells for sample, standard and blank wells on the plate.

3. Add 10 μL of target DNA (sample and standard solutions) as triplicate in the relevant wells on the qPCR plate, plus add a triplicate of blank wells containing 10 μL of water (No template controls).

4. Seal the plate with caps or optically clear film.

5. Centrifuge the plate to get the contents at the bottom.

6. Load the plate on the instrument and run a qPCR reaction machine for 40 cycles (Table 4).

7. In SYBR Green assay, perform a melting curve analysis to confirm the specificity of amplification.

Table 3
Master Mixes for SYBR Green qPCR at 200 nM final of each primer

Master mix contents	Concentration initial	Concentration final (25 μL)	SYBR Green 1 reaction (μL)
Master Mix SYBR Green	2×	1×	12.5
DEPC H$_2$O			1.5
Forward primer	10 μM	200 nM	0.5
Reverse primer	10 μM	200 nM	0.5
Total			15
Sample addition			10

Table 4
Cycling conditions for SYBR Green using StepOnePlus system

Number of cycles	Phase	Temperature (°C)	Acquisition mode	Time
1	*Enzyme activation*			
		95	None	10 min
45	*Amplification*			
	Amplification	95	None	15 s
	Annealing extension	60	Single	60 s
1	*Melting curve (Only in SYBR Green Assay)*			
	Denaturation	95	None	15 s
	Equilibration	65	None	60 s
	Ramp	97	Continuous	0.5–0.9 °C/s

3.5.2 qPCR Validation Before Data Analysis

Once the qPCR reaction is finished, a validation of assay is performed to ensure the quality and reliability of the results.

1. *No template controls*
   It is essential to check that there is no amplification in these tubes to avoid distortion of the results.

2. *Analysis of Melting Curve*
   The analysis of melting or dissociation curve is necessary for nonspecific intercalating dyes assay. Only the samples having a single peak in their dissociation curve can be interpreted.

3. *Analysis of Standard Curve*
   Draw the standard curve and calculate the qPCR efficiency as described above.

3.6 Data Analysis

3.6.1 Normalization of the Data

There are several ways to analyze RT-qPCR data [7]. Here we explain the method normalized relative quantification [8] that is effective and reliable for embryo transcript analysis. This model constituted an improvement over the classic delta-delta-Ct method because it takes account normalization by several reference genes and provides a normalized ratio of expression.

The principle of the delta-Cq quantification method is that a difference (delta) in Cq values between two samples (sample of interest and a calibrator sample) is transformed into a ratio of expression (relative quantification) using the exponential function with the efficiency of the PCR reaction as its base. In theory, any sample can be used as a calibrator; most often a real untreated control is used. In case of gene expression kinetics during embryo development, one specific stage can be used as a calibrator. Such ratios of expression are calculated for the gene of interest (R_{goi}) and the reference genes (R_{refg}).

1. The normalized ratio of expression (R_{Nor}) between the sample to be analyzed and the calibrator is expressed as the ratio of two distinct ratios, i.e., the ratio of the gene of interest in the sample to be analyzed and the calibrator sample (R_{goi}) divided by the ratio of the reference genes in the sample to be analyzed and the calibrator (R_{refg}):

$$R_{Nor} = R_{goi} \, / \, R_{refg}$$

(For the calculation of R_{goi} and R_{refg} *see* **Note 15**).

2. Taking into account the efficiencies of the different PCR reactions, R_{Nor} is calculated according to the formula:

$$R_{Nor} = \frac{\left(1 + E_{goi}\right)^{-\Delta Ct_{goi}}}{\sqrt[f]{\prod_{i \neq 0}^{f}\left(1 + E_{refg_i}\right)^{-\Delta Ct_{refg_i}}}}$$

3. In case of an exogenous transcript reference:

$$R_{Nor} = \frac{\left(1 + E_{goi}\right)^{-\Delta Ct_{goi}}}{\left(1 + E_{refg}\right)^{-\Delta Ct_{refg}}}$$

Software such as qBasePlus software provided by Biogazelle, automatically calculates these ratios starting from the rough RT-qPCR data.

3.6.2 Statistical Analysis

Since the number of PCR replicates for the same sample is often low (minimum recommended three), it is better to use the median value of the PCR triplicates (median of the Cq values). Depending on the number of biological repetitions and on the normal (or non-normal) distribution of the results, either nonparametric (few repetitions, non-normal distribution) or parametric (higher number of biological repetitions, normal distribution) tests have to be used for the statistical analyses of the results.

4 Notes

1. The number of embryos placed in the tube should be verified under binocular loop after transfer because the embryos tend to stick to the pipette.

2. The isolation of RNA is of particular importance since RNA may get degraded or contaminated with genomic DNA, protein and/or organic solvents which may hamper the quality of results produced. The embryonic RNA extraction uses adsorption chromatography. This method does not depend upon the technical operator and reduces solvent traces in the samples. Moreover, RNAs are recovered in a small final volume.

3. If the number of embryos is the same in all samples, it is better to make a mix containing the carrier and the exogenous RNA and distribute it in each sample in order to avoid the pipetting errors. The volume of the mix must not exceed 10 % of the extraction buffer volume.

4. DNase Dilution: to an aliquot of 5 μL DNase I solution stock (2.72 Kunitz Units/μL in water) add 35 μL RDD buffer. Mix by gently inverting the tube or soft pipetting, and centrifuge briefly to collect residual liquid from the sides of the tube. Keep the tube of DNase I on ice before use. DNase I is especially sensitive to physical denaturation. Do not vortex. DNAse I has to be aliquoted and stored at −20 °C to avoid freezing thawing cycles.

5. In RNA isolation, the efficiency of the procedure is affected as some quantity of RNA remains attached to the column filters. Carrier RNA is used to estimate a proportional loss of embryonic RNA. Hence, the embryonic RNA in the sample after isolation does not represent the initial number of embryos. It rather represents an approximate number of embryos which here is called "Equivalent Embryo". The use of "equivalent embryo" as a unit to be used for reverse transcription and qPCR makes a uniformity across different samples.

6. Random primers generate cDNA molecules with a better 3′ to 5′ coverage but less specificity to mRNA, while oligo (dT) primers will specifically target mRNA molecules. In some cases, random primers are preferred particularly when there are chances of mRNA deadenylation, e.g., in oocytes.

7. Negative control without RNA sample is necessary to control the absence of contaminating amplicons in the solutions.

8. In order to get a precise pipetting, we preferably dilute the number of equivalent embryos to be used (determined in a pilot experiment) in 10 µL. The volume of undiluted RT product must not exceed 10 % of the volume of PCR to avoid PCR inhibition by RT buffer salts. It is preferred that a "pilot experiment" should be performed for each gene to be tested, in order to determine the minimum "equivalent embryo" number required in each qPCR reaction to obtain Cq values in the linear range of the standard curve.

9. Nonspecific fluorescent dyes intercalate in between the double stranded DNA molecules and fluoresce significantly more in that state than the free form. Sybr Green is the most commonly used dye for qPCR amplification. In Sybr Green assay, the increase in fluorescence signal is proportional to generation of double stranded DNA during a qPCR reaction.

10. In fluorescent probes system, in addition to the primers, a third sequence specific probe is designed which is complementary to the sequence of the gene to be analyzed and hybridizes in between the two primers. Fluorescent probes consist of short oligonucleotide sequence and are covalently attached to fluorescent dyes. There are several types of fluorescent probes available (reviewed in ref. 1). The sequence specificity of the probes makes this assay much more specific than the intercalating dyes. Furthermore, this assay may be multiplexed. However, this assay is expensive and its optimization is much more complex due to the presence of the probe and possible multiplexing.

11. There are multiple software packages available for primer and probe design such as PrimerExpress (Applied Biosystems®). Moreover, some Web based free access software may also be used like PrimerQuest (IDT®) or Primer3PLus® (Whitehead Institute for Biomedical Research) to design primers.

12. In case of multiple dissociation/melting curves the strategy is either (a) increasing the annealing temperature or (b) decreasing the primer concentration or (c) reducing the number of cycles if a curve corresponding to a secondary product is produced at late cycles.

13. Ideally, a pilot study should be conducted to select the best set of reference genes by testing multiple candidates across all conditions and all embryonic stages to be analyzed.

14. In practice, small variations in the master mix formulation can significantly affect the assay quality. Therefore, it is recommended to use commercial master mixes which are optimized to work in most standards. Moreover, the choice of master mix selected should be compatible with the instrument to be used for the qPCR run. To avoid pipetting errors, we will work on

a sample volume of 10 μL that will add to 15 μL of reaction mixture containing primers, dNTPs, buffer and Taq. The cDNA of each sample will be diluted according to the number of equivalent embryo previously determined and taking into account that the volume of the initial undiluted RT added to the qPCR mix shall not exceed 10 % of the final volume.

15. Formulae for Rgoi and Rrefg.

$$R_{goi} = \left(1 + E_{goi}\right)^{-\Delta Ct_{goi}}$$

$$R_{refg} = \left(1 + E_{refg}\right)^{-\Delta Ct_{refg}} \text{ with: } \Delta Ct = Ct^{unk} - Ct^{cal}$$

Where unk is unknown sample and cal is calibrator.

References

1. Bustin SA (2002) Quantification of mRNA using real-time reverse transcription PCR (RT-PCR): trends and problems. J Mol Endocrinol 29:23–39

2. Ståhlberg A, Håkansson J, Xian X et al (2004) Properties of the reverse transcription reaction in mRNA quantification. Clin Chem 50(3):509

3. Vandesompele J, De Preter K, Pattyn F et al (2002) Accurate normalization of real-time quantitative RT-PCR data by geometric averaging of multiple internal control genes. Genome Biol 3(7):RESEARCH0034

4. Derveaux S, Vandesompele J, Hellemans J (2010) How to do successful gene expression analysis using real-time PCR. Methods 50(4): 227–230

5. Bower NI, Moser RJ, Hill JR et al (2007) Universal reference method for real-time PCR gene expression analysis of preimplantation embryos. Biotechniques 42:199–206

6. Khan DK, Dube D, Gall L et al (2012) Expression of pluripotency master regulators during two key developmental transitions: EGA and early lineage specification in the bovine embryo. PLoS One 7(3):e34110

7. Yuan JS, Reed A, Chen F et al (2006) Statistical analysis of real-time PCR data. BMC Bioinformatics 7:85

8. Hellemans J, Mortier G, De Paepe A et al (2007) qBase relative quantification framework and software for management and automated analysis of real-time quantitative PCR data. Genome Biol 8(2):19

Chapter 15

Studying Bovine Early Embryo Transcriptome by Microarray

Isabelle Dufort, Claude Robert, and Marc-André Sirard

Abstract

Microarrays represent a significant advantage when studying gene expression in early embryo because they allow for a speedy study of a large number of genes even if the sample of interest contains small quantities of genetic material. Here we describe the protocols developed by the EmbryoGENE Network to study the bovine transcriptome in early embryo using a microarray experimental design.

Key words Gene expression, Microarrays, Cow, Embryo, Transcriptome

1 Introduction

In the reproductive biology field, progress in technical advances has evolved at a much faster pace than our understanding of the underlying physiology. As a result, the potential short and long-term impacts of routinely performed assisted reproduction techniques have not been thoroughly assessed. In the case of early embryos, studies seeking to determine the impacts of ART have long been restricted to phenotypical observations [1–6]. A deep knowledge of the phenomenon involved during these early phases of development is crucial to understand the core mechanisms responsible for the onset of life, and to optimize the safety and efficiency of in vitro reproductive technologies. The advents of genomics/transcriptomics, which offer the possibility to amplify genetic material, have enabled scientists to ask critical questions regarding early embryo development [7–10]. It is hypothesized that by comparing a wide array of different treatments, some common grounds could be defined in order to more precisely characterize these impacts. In the past years, extensive research has given rise to new high-throughput technologies for the study of gene expression. So far, microarrays have been the most commonly used high-throughput method in this field. They offer the possibility to

Nathalie Beaujean et al. (eds.), *Nuclear Reprogramming: Methods and Protocols*, Methods in Molecular Biology, vol. 1222, DOI 10.1007/978-1-4939-1594-1_15, © Springer Science+Business Media New York 2015

measure the level of a more-or-less wide number of predetermined transcripts in a given sample. Yet commercially available microarray platforms are not designed to account for potential sample processing biases [11, 12]. In addition, the transcriptome of early embryos greatly differs from that of somatic cells [13]. For example, it is well known that the early embryo subsists through the initial developmental stages by utilizing the maternal supplies provided by the oocyte until such time as its own genome is activated [14–17]. Furthermore, previous works involving the production of subtracted cDNA libraries have highlighted the presence of numerous novel transcribed regions (NTRs) that are rarely represented on commercial platforms [7, 18]. These platforms generate large amounts of data that require the integration of comparative RNA abundance values in the physiological context of early development for their full benefit to be appreciated. The unique characteristics of pre-hatching development, especially in terms of RNA management, combined with the numerous options for sample and data processing, profoundly impact the resulting gene list. It is a fact that microarray data generation requires considerable sample-handling steps, and this is especially true for mammalian oocytes and pre-hatching embryos as they represent a rare source of biological material, thus requiring extensive care in handling and the introduction of a global amplification step. For the establishment of our own microarray platform, we conducted many tests in order to define and control the sources of methodological variations [11]. In this chapter, we describe the protocol developed by the Embryo GENE Network to study early bovine embryo development using microarrays [19]. The EmbryoGENE Bovine microarray is coupled to its own analysis platform, comprising a laboratory Information Management System (LIMS), a microarray quality control module, and microarray analysis software.

2   Materials

2.1   Isolation of Total RNA from Blastocysts

1. ARCTURUS PicoPure RNA Isolation Kit (Life Technologies; The following reagents and materials are supplied from the manufacturer: Conditioning Buffer, Extraction Buffer, Ethanol 70 %, Wash Buffer 1, Wash Buffer 2, Elution Buffer, RNA Purification Columns with collection tubes and microcentrifuge tubes).

2. RNase-Free DNase Set (Qiagen).

3. Agilent 2100 Bioanalyzer (Agilent technologies).

4. RNA 6000 Pico total RNA Kit (Agilent technologies); The following reagents and materials are supplied from the manufacturer: RNA 6000 Pico Chips, Pico Dye Concentrate, Pico

Marker, Pico Conditioning Solution, Pico Gel Matrix, Pico Ladder, Spin Filters Tubes for Gel-Dye Mix, Safe-Lock Eppendorf Tubes PCR clean (DNase/RNase free).

2.2 RNA Amplification

1. Arcturus™ RiboAmp® HS PLUS Amplification Kit (Life Technologies; The following reagents and materials are supplied from the manufacturer: first Strand Master Mix, first Strand Enzyme Mix, Enhancer, Nuclease Mix, second Strand Master Mix, second Strand Enzyme Mix, Primer 1, Primer 2, Primer 3, Control RNA, IVT Buffer, IVT Master Mix, IVT Enzyme Mix, DNase Mix, amplification purification kit for DNA and RNA).

2. Superscript III (200 U/μl) (Life Technologies).

3. Nuclease-Free Water.

4. NanoDrop ND-1000 Spectrophotometer (Thermo Fisher Scientific).

2.3 Labelling of aRNA

1. ULS™ Fluorescent Labeling Kit for Agilent arrays (with Cy3 and Cy5) (Kreatech, The following reagents and materials are supplied from the manufacturer: Labeling solution, Cy3-ULS, Cy5-ULS).

2. Nuclease-Free Water.

3. NanoDrop ND-1000 Spectrophotometer (Thermo Fisher Scientific).

4. ARCTURUS PicoPure RNA Isolation Kit (Life Technologies; *see* Subheading 2.2, **step 1**).

2.4 Hybridization

1. RNA Spike-In Kit, Two-Color (Agilent Technologies; *see* **Note 1**).

2. Nuclease-Free Water.

3. Gene Expression Hybridization Kits (Agilent Technologies; The following reagents and materials are supplied from the manufacturer: 2× Hi-RPM Hybridization Buffer, 25× Fragmentation Buffer, 10× Gene Expression Blocking Agent).

4. Hybridization Chamber Kit—SureHyb enabled, Stainless (Agilent Technologies).

5. Hybridization oven rotator.

6. Hybridization Gasket Slide Kit (100)—four microarrays per slide format (Agilent Technologies).

7. Gene Expression Wash Buffer Kit containing Wash Buffer 1, Wash Buffer 2, and Triton X-102 (Agilent Technologies). Before use, add 2.0 ml of Triton X-102 to each 4 L containers of Wash Buffer 1 and 2. Mix well.

8. Acetonitrile.

9. Stabilization and Drying Solution (Agilent technologies; *see* **Note 2**).

10. Slide-staining dish, with slide rack.

2.5 Array Scanning

1. PowerScanner (Tecan).

2. ArrayPro software 6.3 (MediaCybernetics).

3 Methods

3.1 Experimental Design

The experimental design involved two color hybridizations with full dye-swap technical replication. For all biological contrasts, in vivo vs. in vitro, a simple direct comparison was performed with four independent biological replicates (each composed of a pool of ten blastocysts) per treatment.

3.2 Isolation of Total RNA from In Vivo and In Vitro Produced Blastocysts

1. Day-7 blastocysts produced in vitro [20] and in vivo blastocysts [21] are collected.

2. Immediately after collection, the blastocysts are washed three times in PBS 1x, pooled by ten and frozen at −80 °C in 20 μl of PBS 1x until total RNA extraction.

3. Prepare total RNA using Arcturus picopure RNA isolation kit (Life technologies). This kit is specially design to recover total cellular RNA from pico-scale samples.

4. Precondition the column by pipetting 250 μl Conditioning Buffer onto the purification column filter membrane.

5. Incubate the RNA Purification Column with Conditioning Buffer for 5 min at room temperature.

6. Centrifuge the purification column in the provided collection tube at $16,000 \times g$ for 1 min.

7. Evaluate the amount of PBS present in the tube with the sample (must not exceed 30 μl).

8. Add 100 μl of extraction buffer to the sample, gently mix, spin down, and incubate 30 min at 42 °C.

9. Add 100 μl of 70 % EtOH plus the corresponding amount of liquid present in the tube at **step 5** (ex: 100 μl of 70 % EtOH + 20 μl corresponding to the 20 μl of PBS 1x present with the blastocysts).

10. Mix well by pipetting up and down and transfer the liquid in a preconditioned column.

11. The Dnase 1 treatment is done directly on the column using 10 μl of DNase 1 (Qiagen) and 30 μl of RDI buffer to eliminate genomic DNA contamination.

12. Incubate the columns 10 min at room temperature.

13. Add 40 μl of Wash Buffer 1 and centrifuge 15 s at 8,000×*g*.

14. Add 100 μl of Wash Buffer 2, centrifuge for 1 min at 8,000×*g*.

15. Add 100 μl of Wash Buffer 2, centrifuge for 2 min at 16,000×*g*.

16. Turn the column (*see* **Note 3**) in the centrifuge at a 180° angle and re-centrifuge for 1 min at 16,000×*g*.

17. Transfer the purification column to a new 0.5 ml microcentrifuge tube provided in the kit. Make sure that no liquid remains in the column or at the base of the column.

18. Pipette 13.5 μl of Elution Buffer directly onto the membrane of the purification column (Gently touch the tip of the pipette to the surface of the membrane while dispensing the elution buffer to ensure maximum absorption of EB into the membrane).

19. Incubate the purification column for 1 min at room temperature.

20. Centrifuge the column for 1 min at 1,000×*g* to distribute EB in the column, then for 1 min at 16,000×*g* to elute the labelled aRNA.

21. The elution is done in 13.5 μl of Elution buffer. The extraction column normally retained 1.0 μl of liquid. 10 μl is conserved at −80 °C until amplification while 1.5 μl is used for quantification and quality (*see* **Note 4**) check using the Bioanalyzer (Agilent).

3.3 RNA Amplification (See Note 5)

1. Prepare the amplified anti-sense RNA using the RNA Amplification HS (Life technologies) following the manufacturer recommendations.

2. Before the elution, add an additional wash using 200 μl of 80 % EtOH. Centrifuge for 2 min at 16,000×*g*. The final material is eluted in 30 μl of elution buffer (*see* **Note 6**).

3. The amplified materiel is quantified using the NanoDrop and the purity is verified with the 260/280 and 260/230 ratio (*see* **Note 7**).

4. A typical RNA amplification from a pool of ten day-7 blastocysts generally yields between 60 and 80 μg of aRNA.

5. Quality of the amplification can also be checked using the Bioanalyzer and a nano chip. The resulting aRNA has a profile ranging from 200 to 600 pb (*see* **Note 8** and Fig. 1).

3.4 Labelling of aRNA (See Note 9)

1. Transfer 2 μg (*see* **Note 10**) of aRNA to RNase-free microfuge tube.

2. Add 2 μl of ULS-cye3 or cye5 and 2 μl of labelling buffer.

3. Adjust to a final volume of 20 μl with RNAse-free water.

4. Mix gently and incubate the tubes in a thermocycler at 85 °C for 15 min.

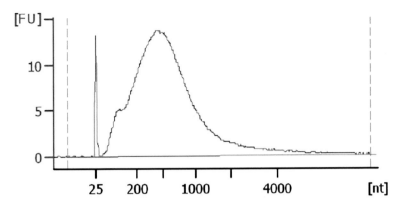

Fig. 1 Profile of aRNA after a two-round amplification on the bioanalyzer

5. Place samples on ice for at least 1 min, spin down to collect content of tube before proceeding with purification.

6. For cleanup of the cy-dye-labelled aRNA, use the Picopure RNA extraction kit (Life Technologies).

7. Precondition the column by pipetting 250 μl Conditioning Buffer onto the purification column filter membrane.

8. Incubate the RNA Purification Column with Conditioning Buffer for 5 min at room temperature.

9. Centrifuge the purification column in the provided collection tube at $16,000 \times g$ for 1 min.

10. Add 100 μl of Extraction buffer to the labelled aRNA reaction and mix well by pipetting up and down.

11. Add 120 μl EtOH 70 %, mix well by pipetting up and down and transfer the 240 μl to a preconditioned purification column.

12. To bind RNA to the column, centrifuge for 2 min at $100 \times g$ followed by a centrifugation at $16,000 \times g$ for 30 s to remove flowthrough.

13. Add 100 μL Wash Buffer 1 into the purification column and centrifuge for 1 min at $8,000 \times g$.

14. Add 100 μL Wash Buffer 2 into the purification column and centrifuge for 1 min at $8,000 \times g$.

15. Add another 100 μl Wash Buffer into the purification column and centrifuge for 2 min at $16,000 \times g$.

16. Turn the column (*see* **Note 3**) in the centrifuge at a 180° angle and re-centrifuge for 1 min at $16,000 \times g$.

17. Transfer the purification column to a new 0.5 ml microcentrifuge tube provided in the kit. Make sure that no liquid remains in the column or at the base of the column.

18. Pipette 15 µl of Elution Buffer directly onto the membrane of the purification column (Gently touch the tip of the pipette to the surface of the membrane while dispensing the elution buffer to ensure maximum absorption of EB into the membrane).

19. Incubate the purification column for 1 min at room temperature.

20. Centrifuge the column for 1 min at $1,000 \times g$ to distribute EB in the column, then for 1 min at $16,000 \times g$ to elute the labelled aRNA.

21. Check the concentration of the labelled aRNA by measuring absorbance and dye concentration using a NanoDrop.

22. Determine the degree of labelling (DoL) as follows: measure absorbance at 260 and 555 nm (for CY3) or 647 nm (for CY5) using a NanoDrop. The result (ng/µl) at 260 nm indicates aRNA concentration and the other absorbance indicates the amount of labelling (ng/µl). You should have between 30 and 50 pmol/µg of labelling to have a good hybridization.

3.5   Hybridization

1. Get the needed quantity of chamber assemblies and packaged gasket slides to process samples.

2. Prepare the Agilent hybridization chambers, gaskets, and slides.

3. The labelled target and hybridization solution will be loaded onto the gasket slide, not the microarray slide.

4. For each array to be hybridized, prepare a tube containing 825 ng of cy3-labelled-aRNA, 825 ng of cy5-labelled-aRNA, 2.75 µl of Agilent spikes (*see* **Note 1**), 11.0 µl blocking agent, 2.2 µl of fragmentation buffer.

5. Complete the volume to 55 µl using RNase-free water.

6. Incubate at 60 °C for exactly 15 min to fragment RNA (*see* **Note 11**).

7. Immediately cool on ice for 1 min.

8. Add 2× GEx Hybridization Buffer HI-RPM to the 4 x 44 K array formats at the appropriate volume to stop the fragmentation reaction.

9. Mix well by careful pipetting. Take care to avoid introducing bubbles. Do not mix on a vortex mixer; mixing on a vortex mixer introduces bubbles.

10. Spin for 1 min at room temperature at $13,000 \times g$ in a microcentrifuge to drive the sample off the walls and lid and to aid in bubble reduction.

11. Use immediately. Do not store.

12. Place sample on ice.

13. Load 100 μl of the hybridization cocktail on the gasket slide. When the four gaskets area are loaded with the hybridization cocktails, with gloved fingers, grab a microarray from its slide holder between your thumb and index finger, numeric barcode side facing up and Agilent label facing down, and carefully lower it on top of the gasket slide. Barcode ends of both slides must line up at rectangular end of the chamber base.

14. Close the hybridization chamber.

15. Incubate the arrays for 17 h at 65 °C with 10 rpm rotation in the hybridization oven.

16. Incubated Wash Buffer 2 overnight at 42 °C.

17. Prepare five slide-staining dishes (*see* **Note 12**) for slide washes each containing Wash Buffer 1 at room temperature, Wash Buffer 2 at 42 °C (*see* **Note 13**), acetonitrile and stabilization solution (*see* **Note 2**) at room temperature.

18. After 17 h of hybridization, remove the gasket slide from the array in Wash Buffer 1 at room temperature.

19. Incubate the arrays in Wash Buffer 1 for 3 min at room temperature.

20. Transfer the slide to the container with Wash Buffer 2 at 42 °C and incubate at room temperature for 3 min.

21. Transfer the slides to the acetonitrile for 10 s.

22. Transfer the slides to the stabilization solution for 30 s.

23. Slowly take out the slides (*see* **Note 14**) and let dry for 30 s before scanning.

3.6 Array Scanning

1. Make sure that no liquid remains at the surface of the array slide before the scan.

2. Scan the slide using Powerscanner (Tecan) at 5 μm using specific wavelength excitation and emission for cy3 and cy5 (*see* **Note 15**).

3.7 Computational Analyses of RNA Expression

1. Once the .tiff images have been produced, you can extract the data using specific software ArrayPro 6.3.

2. The raw data and characteristics of the experiments are then processed through the EmbryoGENE Laboratory Information Management System (LIMS) ELMA (http://elma.embryo-gene.ca).

3. Raw data are analyzed using the Flexarray 1.6.1 software package for statistical analysis and visualization of microarray expression data (Michal Blazejczyk, Mathieu Miron, Robert Nadon (2007), FlexArray: statistical data analysis software for gene expression microarrays. Genome Quebec, Montreal, Canada, URL: http://genomequebec.mcgill.ca/FlexArray).

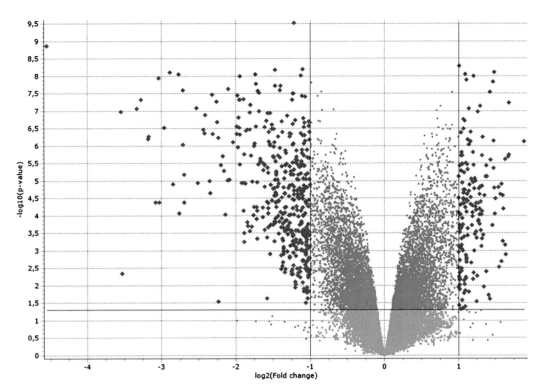

Fig. 2 Volcano plot of expression for in vivo and in vitro blastocysts

4. Data are first transformed using simple background correction, then, a Loess normalization within array was performed, followed by a quantile between arrays normalization.

5. Statistical comparison between treatments (in vivo vs. in vitro) is done using the Limma algorithm. Significant differences between treatments are considered significant with a fold change >2 and a P value < 0.05. Visualization of the results can be made using a volcano plot (Fig. 2).

4   Notes

1. Spikes for quality control of the hybridization are prepared using the RNA Spike-In Kit, Two-Color (Agilent Technologies). Those spikes (A&B) are normally added to the RNA before amplification but here they are used has a quality control for the hybridization. Both spikes have been prepared according to the manufacturer instruction and then submitted to two rounds of RNA amplification using the Arcturus™ RiboAmp® HS PLUS Amplification Kit (Life Technologies). Spike A was labelled with cy3 and spike B labelled with cy5. Both labelled spikes were mixed together in equal amounts and diluted to a final concentration of 500 pg/μl.

2. The Stabilization and drying solution often forms a precipitate. Make sure that all the precipitate is well dissolved by heating it at a maximum of 40 °C. It may take several hours.

3. Before the elution, the columns need to be very dry to eliminate the transfer of any Wash Buffer into the eluate. Spinning the column for another minute will get rid of any left-over buffer. Care must also be taken to avoid liquid splashing when the tubes are manipulated.

4. The success of amplification using the RiboAmp HS^{Plus} RNA Amplification Kit depends on the quality of the source RNA. The Quality of the RNA is very important in microarray analysis and samples with an RNA integrity number (RIN) lower than 7.5 should not be considered for hybridization. Note: When using pre-hatched embryos, especially RNA prepared from oocytes and embryos before the maternal-to-embryonic transition, the RIN of the samples is often lower than 7 even if the RNA is of good quality due to a variation in the 28S/18S rRNA ratio [22].

5. Using isolated total RNA, rather than mRNA, for amplification is recommended to reduce the loss of valuable transcripts during mRNA isolation from extremely small samples. The RNA must be provided in a water based buffer, without the presence of organic solvents, salts, or contaminating cellular material. It is not recommended to use phenol based extraction techniques to avoid inhibiting the amplification reaction.

6. The procedure consists of reverse transcription with an oligo(dT) primer bearing a T7 promoter and in vitro transcription of the resulting DNA with T7 polymerase to generate hundreds to thousands of antisense RNA copies of each mRNA in a sample. The procedure is repeated for a second round to produce more aRNA. The advantage of that amplification is that on top of being linear, it also produces anti-sense RNA that can be directly hybridized on oligo array without the need to transform it to cDNA.

7. Quality criteria:

   $OD_{260}/OD_{280} > 1.8$ (<1.8 presence of proteins)

   $OD_{260}/OD_{230} > 1.8$ (<1.8 presence of salts, organic chemicals, etc.)

   If the ratios are lower than 1.8, it is recommended to proceed with a purification method of the RNA before labelling.

8. After the two-round amplification, aRNA fragments are expected to range between 200 and 800 pb and have a profile similar to the one shown in Fig. 2.

9. When working with cye dye always be careful of the degree of ozone present in your environment since the cy5 dye is sensitive

to ozone degradation. Laboratory ozone levels should be kept below 2 ppb to achieve high quality microarray data [23].

10. The recommended amount of aRNA to be labelled with the ULS™ aRNA Labeling Kit for Agilent (Kreatech) is 2 μg but you can go up to 4 μg without affecting the efficacy of labelling. Over that amount, the pmol/μg of labelling will fall under the acceptable number for hybridization.

11. The fragmentation is made over a 15 min period instead of the 30 min recommended by the manufacturer. The aRNA produced after a two-round amplification is shorter than the one produced with only one round so the time of fragmentation needs to be shorten to make sure that fragments are not too short or degraded.

12. Before using the Slide-staining dishes for microarrays washes, rinse the dishes with acetonitrile and water. The presence of other fluorescent compound or detergent may alter the quality of the microarray.

13. Washes are performed at room temperature with a solution previously heated up at 42 °C.

14. Always take out the slide slowly from the stabilization and drying solution so that no marks or lines will appear on the slide.

15. The slides are scanned using the autogain function on the Powerscanner (Tecan).

Acknowledgment

This work was supported by Natural Science and Engineering Research Council of Canada Strategic Network Research Grant NETPG 340825-06. The authors thank L'Alliance Boviteq, Saint-Hyacinthe, Québec, Canada and Julie Nieminen for her help in editing the final manuscript.

References

1. Alexopoulos NI et al (2008) Developmental disparity between in vitro-produced and somatic cell nuclear transfer bovine days 14 and 21 embryos: implications for embryonic loss. (Translated from eng). Reproduction 136(4): 433–445

2. Crosier AE, Farin PW, Dykstra MJ, Alexander JE, Farin CE (2001) Ultrastructural morphometry of bovine blastocysts produced in vivo or in vitro. (Translated from eng). Biol Reprod 64(5):1375–1385

3. McEvoy TG, Ashworth CJ, Rooke JA, Sinclair KD (2003) Consequences of manipulating gametes and embryos of ruminant species. (Translated from eng). Reprod Suppl 61: 167–182

4. McEvoy TG, Sinclair KD, Young LE, Wilmut I, Robinson JJ (2000) Large offspring syndrome and other consequences of ruminant embryo culture in vitro: relevance to blastocyst culture in human ART. (Translated from Eng). Hum Fertil (Camb) 3(4):238–246

5. Sinclair KD et al (1999) Aberrant fetal growth and development after in vitro culture of sheep zygotes. (Translated from eng). J Reprod Fertil 116(1):177–186

6. van Wagtendonk-de Leeuw AM et al (2000) Effects of different reproduction techniques: AI MOET or IVP, on health and welfare of bovine offspring. (Translated from eng). Theriogenology 53(2):575–597

7. Bui LC et al (2009) Retrotransposon expression as a defining event of genome reprogramming in fertilized and cloned bovine embryos. (Translated from eng). Reproduction 138(2): 289–299

8. Hansen PJ, Block J, Loureiro B, Bonilla L, Hendricks KE (2010) Effects of gamete source and culture conditions on the competence of in vitro-produced embryos for post-transfer survival in cattle. (Translated from eng). Reprod Fertil Dev 22(1):59–66

9. Nowak-Imialek M et al (2008) Messenger RNA expression patterns of histone-associated genes in bovine preimplantation embryos derived from different origins. (Translated from eng). Mol Reprod Dev 75(5):731–743

10. Rizos D et al (2003) Bovine embryo culture in the presence or absence of serum: implications for blastocyst development, cryotolerance, and messenger RNA expression. (Translated from eng). Biol Reprod 68(1):236–243

11. Gilbert I et al (2010) Providing a stable methodological basis for comparing transcript abundance of developing embryos using microarrays. (Translated from eng). Mol Hum Reprod 16(8):601–616

12. Robert C (2010) Microarray analysis of gene expression during early development: a cautionary overview. (Translated from eng). Reproduction 140(6):787–801

13. Vallee M et al (2009) Revealing the bovine embryo transcript profiles during early in vivo embryonic development. (Translated from eng). Reproduction 138(1):95–105

14. Bachvarova R, De Leon V, Johnson A, Kaplan G, Paynton BV (1985) Changes in total RNA, polyadenylated RNA, and actin mRNA during meiotic maturation of mouse oocytes. (Translated from eng). Dev Biol 108(2): 325–331

15. Bachvarova RF (1992) A maternal tail of poly(A): the long and the short of it. (Translated from eng). Cell 69(6):895–897

16. De Leon V, Johnson A, Bachvarova R (1983) Half-lives and relative amounts of stored and polysomal ribosomes and poly(A)+RNA in mouse oocytes. (Translated from eng). Dev Biol 98(2):400–408

17. Paynton BV, Bachvarova R (1994) Polyadenylation and deadenylation of maternal mRNAs during oocyte growth and maturation in the mouse. (Translated from eng). Mol Reprod Dev 37(2):172–180

18. Robert C, Barnes FL, Hue I, Sirard MA (2000) Subtractive hybridization used to identify mRNA associated with the maturation of bovine oocytes. (Translated from eng). Mol Reprod Dev 57(2):167–175

19. Robert C et al (2011) Combining resources to obtain a comprehensive survey of the bovine embryo transcriptome through deep sequencing and microarrays. (Translated from eng). Mol Reprod Dev 78(9):651–664

20. Cagnone GL, Dufort I, Vigneault C, Sirard MA (2012) Differential gene expression profile in bovine blastocysts resulting from hyperglycemia exposure during early cleavage stages. (Translated from eng). Biol Reprod 86(2):50

21. Plourde D et al (2012) Contribution of oocyte source and culture conditions to phenotypic and transcriptomic variation in commercially produced bovine blastocysts. (Translated from eng). Theriogenology 78(1): 116–131, e111–113

22. Gilbert I et al (2009) The dynamics of gene products fluctuation during bovine prehatching development. (Translated from eng). Mol Reprod Dev 76(8):762–772

23. Fare TL et al (2003) Effects of atmospheric ozone on microarray data quality. (Translated from eng). Anal Chem 75(17):4672–4675

Chapter 16

Methylation of Specific Regions: Bisulfite-Sequencing at the Single Oocyte or 2-Cell Embryo Level

Annick Lefèvre and Thierry Blachère

Abstract

To question the possible implication of an alteration of the DNA methylation of imprinted genes in normal development failure observed following fertilization in ART centers, it has been necessary to develop a reproducible and highly efficient method to perform analysis at the one cell level. We have thus developed a very efficient protocol for methylation studies on individual oocytes or cleavage-stage embryos. All the different steps were optimized, from DNA extraction, to limit DNA degradation and give a high success rate of bisulfite converted DNA, to amplification of the bisulfite modified DNA.

Key words Oocyte, DNA methylation, PCR at the single cell level

1 Introduction

The epigenetic status of oocytes and early cleavage-stage embryos (ECSE), particularly the DNA methylation profile of imprinting genes, has been questioned this past decade to investigate on the possible causes of normal development failure following fertilization in ART centers. Human oocytes and early cleavage-stage embryos are parsimoniously available on large scale and are precious samples, making critical the development of a highly efficient and reproducible method to perform analysis at the one cell level. The development of such an individual assay is of particular interest to relate alterations observed in an embryo with those in the oocytes or the sperm of the couple.

The first information on DNA methylation in oocytes was obtained using Southern blot hybridization techniques with DNA cleaved with methylation-sensitive restriction enzymes [1]. This method was very restrictive, since only a limited proportion of potentially methylated sites can be analyzed, and needs quite high quantities of DNA. More recently, a different approach has been developed, based on the ability of sodium bisulfite to convert specifically unmethylated cytosine residues to uracil in single-stranded DNA.

Nathalie Beaujean et al. (eds.), *Nuclear Reprogramming: Methods and Protocols*, Methods in Molecular Biology, vol. 1222, DOI 10.1007/978-1-4939-1594-1_16, © Springer Science+Business Media New York 2015

The modified DNA is thus amplified by PCR with a pair of specific primers, yielding a double strand fragment in which all uracil and thymidine residues have been amplified as thymidine and only initially methylated cytosine residues (5-MeC) have been amplified as cytosine. Following sequencing, this method enables the DNA methylation identification at the individual nucleotide level, being thus one of the most powerful methods for DNA methylation analysis [2].

The challenge when analysis is performed on an individual oocyte or early cleavage-stage embryo is to protect a minute amount of genomic DNA from degradation, and subsequent loss during the numerous protocol steps. There are a number of steps that are critical and required special care and optimization. The first one is to obtain oocytes and ECSE free from cumulus cells. Before removal of the zona pellucida, cumulus cells must be carefully eliminated by adequate treatments followed by several washing and careful shaking under an inverted microscope with Hoffman modulation contrast optics. Zona pellucida denuded oocytes are very adhesive and may stick any remaining contaminating cell. All steps have to be optimized to limit DNA lost and degradation. Special care is taken for the design of PCR primers that must unable amplification from a few gene copies. We have developed a very efficient method for methylation studies on individual oocytes or cleavage-stage embryos [3–5], which gives a high success rate of bisulfite converted DNA.

2   Materials

The use of kits, Taq polymerase, or others components from designated companies was tested and is recommended in order to obtain results taking account of the sample nature and of the very the low DNA quantity available.

2.1 Sample Preparation

1. Proteinase K: resuspended in PBS, 9 U/ml.
2. Hyaluronidase solution: 150 U/ml.

2.2 Bisulfite Conversion and Cleanup of Bisulfite Converted DNA

1. EpiTect Plus Lyse All Bisulfite Kit, Qiagen for 48 preps.

   Buffer EL 25 ml.

   Lysis Buffer FTB 0.8 ml.

   Proteinase K 1.4 ml.

   Bisulfite Mix (aliquots for eight reactions) six tubes.

   DNA Protect Buffer 1.9 ml.

   RNase-Free Water 3 × 1.9 ml.

   MinElute DNA spin columns 48.

   Collection Tubes (2 ml) 96.

Buffer BL 31 ml.

Buffer BW (concentrate; 2×13 ml): add 30 ml ethanol (96–100 %) to Buffer BW and store at room temperature (15–25 °C). Invert the bottle several times before starting the procedure.

Buffer BD (concentrate; 3 ml): add 27 ml ethanol (96–100 %) to Buffer BD and store at 2–8 °C. Invert the bottle several times before starting the procedure and make sure to close the bottle immediately after use. White precipitates may form in the Buffer BD–ethanol mix after some storage time. These precipitates will not affect the performance of Buffer BD. However, avoid transferring precipitates to the MinElute DNA spin column.

Buffer EB 15 ml.

Carrier RNA (310 μg): add 310 μl RNase-free water to the lyophilized carrier RNA to obtain at 1 μg/μl solution. Dissolve the carrier RNA thoroughly by vortexing. When processing 48 samples in parallel, add the complete volume of dissolved carrier RNA to the bottle of Buffer BL, and check the box on the bottle lid label. If processing fewer samples, split the dissolved carrier RNA into conveniently sized aliquots (e.g., 50 μl) and store at –20 °C. Aliquots can be stored for up to 1 year. If fewer than 48 conversions will be performed in a 2-week period, only make enough Buffer BL–carrier RNA solution as required. Carrier RNA enhances binding of DNA to the spin column membrane, especially if there are very few target molecules in the sample. Carrier RNA is not necessary if >100 ng DNA is used. Add dissolved carrier RNA to Buffer BL. Calculate the volume of Buffer BL and dissolved carrier RNA required for the number of samples to be processed. If Buffer BL contains precipitates, dissolve by heating (maximum 70 °C) with gentle agitation.

2. Ethanol (molecular biology grade, 96–100 %).

3. Pipets and pipet tips with aerosol barriers for preventing cross-contamination.

4. 0.2 ml reaction tubes or 8-well strips.

5. Thermal cycler with heated lid. Since the bisulfite reaction is not overlaid with mineral oil, only thermal cyclers with heated lids are suitable for this procedure.

6. 1.5 ml microcentrifuge tubes for elution steps.

7. Microcentrifuge.

8. Optional: Heating block, thermomixer, or heated orbital incubator.

9. MinElute DNA spin columns.

2.3 PCR Amplification of Bisulfite Converted DNA

Characteristics of Taq Polymerase

HOT BIOAmp DNA® polymerase (BIOfidal) is a chemically modified BIOAmp DNA® polymerase. At room temperature it is inactive, having no polymerization activity. HOT BIOAmp DNA® polymerase is inactivated by a 15 min incubation step at 95 °C. This prevents extension of nonspecifically annealed primers and primer-dimers formed at low temperature during PCR setup. The enzyme has $5' \rightarrow 3'$ polymerization-dependent exonuclease replacement activity but lacks $3' \rightarrow 5'$ exonuclease activity. HOT BIOAmp DNA® polymerase is purified from an *E. coli* strain that carries an overproducing plasmid containing a modified gene of *Thermus aquaticus* DNA polymerase. This enzyme is very stable at high T°, so it can be shipped at room T° and also stored at room T° for long time.

2.4 Extraction and Purification of Amplified Bisulfite Converted DNA

1. E.Z.N.A. Gel Extraction Kit from OMEGA bio-tek.

   HiBind DNA Mini columns.

   2 ml collection tubes.

   Binding buffer (XP2).

   Elution buffer.

   SPW wash buffer to be diluted before use with 100 % ethanol and stored at room temperature.

   Heat block or water bath capable of 60 °C.

   Microcentrifuge capable of at least $13,000 \times g$.

   Vortex.

   Nuclease-free 1.5 ml microcentrifuge tubes.

2.5 Ligation of Amplified Bisulfite Converted DNA to pGEM-T Easy Vector System

1. 1.2 μg pGEM®-T Vector (50 ng/μl), Promega.
2. 12 μl Control Insert DNA (4 ng/μl).
3. 100 U T4 DNA Ligase.
4. 200 μl 2× Rapid Ligation Buffer, T4 DNA Ligase. 60 mM Tris–HCl (pH 7.8), 20 mM MgCl₂, 20 Mm DTT, 2 mM ATP, 10 % polyethylene glycol (MW8000, ACS Grade). Store in single-use aliquots at –20 °C. Avoid multiple freeze–thaw cycles.

2.6 Transformation Protocol

1. JM109 Competent Cells, High Efficiency (6×200 μl), Promega. The genotype of JM109 is recA1, endA1, gyrA96, thi, hsdR17 (rK–, mK+), relA1, supE44, Δ(lac-proAB), (F′, traD36, proAB, lacIqZΔM15). The selection for transformants should be on LB/ampicillin/IPTG/X-Gal plates. For best results, do not use plates that are more than 1 month old.

   Store the Competent Cells at –70 °C. Store all other components at –20 °C.

2. Buffers and solutions:

– IPTG stock solution (0.1 M): 1.2 g IPTG. Add water to 50 ml final volume. Filter-sterilize and store at 4 °C.

– X-Gal solution (2 ml): 100 mg 5-bromo-4-chloro-3-indolyl-β-D-galactoside. Dissolve in 2 ml N,N'-dimethylformamide. Cover with aluminum foil and store at –20 °C.

– LB medium (per liter): 10 g Bacto®-Tryptone; 5 g Bacto®-Yeast Extract; 5 g NaCl. Adjust pH to 7.0 with NaOH.

– LB plates with ampicillin: add 15 g agar to 1 l of LB medium. Autoclave it. Allow the medium to cool to 50 °C before adding ampicillin to a final concentration of 100 µg/ml. Pour 30–35 ml of medium into 85 mm petri dishes. Let the agar harden. Store at 4 °C for up to 1 month or at room temperature for up to 1 week.

– LB plates with ampicillin/IPTG/X-Gal: make the LB plates with ampicillin as above; then supplement with 0.5 mM IPTG and 80 µg/ml X-Gal and pour the plates. Alternatively, 100 µl of 100 mM IPTG and 20 µl of 50 mg/ml X-Gal may be spread over the surface of an LBampicillin plate and allowed to absorb.

– SOC medium (100 ml): 2.0 g Bacto®-Tryptone, 0.5 g Bacto®-Yeast Extract, 1 ml 1 M NaCl, 0.25 ml 1 M KCl, 1 ml 2 M Mg^{2+} stock, filter-sterilized, 1 ml 2 M glucose, filter-sterilized. Add Bacto®-Tryptone, Bacto®-Yeast Extract, NaCl, and KCl to 97 ml distilled water. Stir to dissolve. Autoclave and cool to room temperature. Add 2 M Mg^{2+} stock and 2 M glucose, each to a final concentration of 20 mM. Bring to 100 ml with sterile, distilled water. The final pH should be 7.0.

– 2 M Mg^{2+} stock: 20.33 g $MgCl_2 \cdot 6H_2O$, 24.65 g $MgSO_4 \cdot 7H_2O$. Add distilled water to 100 ml. Filter-sterilize.

– TYP broth (per liter): 16 g Bacto®-Tryptone, 16 g Bacto®-Yeast Extract, 5 g NaCl, 2.5 g K_2HPO_4.

3   Methods

3.1   Sample Preparation

1. Oocytes and ECSE are collected.

2. Denude oocyte/embryo of cumulus cells by repeated pipetting in a hyaluronidase solution.

3. Wash each oocyte three times in fresh PBS.

4. Place oocytes and ECSE in a drop of proteinase K solution for 1–2 min at room temperature to remove zona pellucida and any remaining somatic cells.

5. Wash carefully zona free oocytes and ECSE by transfer successively to three drops of fresh PBS. It is recommended to change the pipet after each transfer to a new drop of PBS.

6. Examination under an inverted microscope with Hoffman Modulation Contrast optics (Leica DM IRB) and selection only of cumulus-free oocytes and embryos for analysis.

7. Placed each oocytes/embryos in 10 µl PBS in a 200 µl reaction tube and saved at −80 °C until bisulfite treatment.

3.2 General Considerations About the Bisulfite Treatment

Incubation of the target DNA with sodium bisulfite results in conversion of unmethylated cytosine residues into uracil, leaving the methylated cytosines unchanged. Therefore, bisulfite treatment gives rise to different DNA sequences for methylated and unmethylated DNA. The most critical step for correct determination of a methylation pattern is the complete conversion of unmethylated cytosines. This is achieved by incubating the DNA in high bisulfite salt concentrations at high temperature and low pH. These harsh conditions usually lead to a high degree of DNA fragmentation and subsequent loss of DNA during purification. Purification is necessary to remove bisulfite salts and chemicals used in the conversion process that inhibit sequencing. Common bisulfite procedures usually require high amounts of input DNA to compensate for DNA degradation during conversion and DNA loss during purification that often lead to low DNA yield, highly fragmented DNA, and irreproducible conversion rates. We use the EpiTect Plus LyseAll Bisulfite Kit according to the manufacturer recommendations. This kit now provides a fast and streamlined procedure for efficient conversion and purification of DNA prepared from oocytes or early stage-cleavage embryos. The kit contains preparation buffers that make it unnecessary to isolate the DNA prior to bisulfite treatment. DNA fragmentation is prevented during the bisulfite conversion by the unique DNA Protect Buffer, which contains tetrahydrofurfuryl alcohol and a pH-indicator dye as a mixing control in reaction setup, allowing confirmation of the correct pH for cytosine conversion.

Furthermore, the bisulfite thermal cycling program provides an optimized series of incubation steps necessary for thermal DNA denaturation and subsequent sulfonation and cytosine deamination, enabling high cytosine conversion rates of over 99 %. Desulfonation, the final step in chemical conversion of cytosines, is achieved by a convenient on-column step included in the purification procedure. The final elution volume can be as low as 10 µl though this may result in a yield reduction.

3.3 Genomic DNA Extraction

For each oocyte/embryo previously collected in individual tube

1. Add 10 µl distilled water, 15 µl Lysis Buffer FTB, and 5 µl proteinase K.

The master mix containing distilled water, Lysis Buffer FTB, and proteinase K may be prepared in advance.

2. Vortex and briefly centrifuge the samples.

3. Incubate samples for 30 min at 56 °C.

4. Proceed as soon as possible with bisulfite conversion.

3.4 Bisulfite Conversion of Genomic DNA

1. Prepare the Bisulfite solution. Dissolve the required number of aliquots of Bisulfite Mix by adding 800 μl RNase-free water to each aliquot. Vortex until the Bisulfite Mix is completely dissolved. This can take up to 5 min (*see* **Note 1**).

2. Add each component directly to the 0.2 ml tube containing an isolated oocyte/ECSE, in this order: a 40 μl final volume of DNA solution in RNase free water, 85 μl of Bisulfite Mix, 15 μl of DNA protect Buffer (140 μl final volume).

3. Close the PCR tubes and mix the bisulfite reactions thoroughly.

4. Store the tubes at room temperature (15–25 °C). DNA Protect Buffer should turn from green to blue after addition to DNA–Bisulfite mix, indicating sufficient mixing and correct pH for the bisulfite conversion reaction.

5. Place the PCR tubes containing the bisulfite reactions into the thermal cycler. Perform the bisulfite DNA conversion using a thermal cycler with a defined thermal cycler program (denaturation, 5 min. at 95 °C; incubation, 25 min at 60 °C; denaturation, 5 min at 95 °C; incubation, 85 min at 60 °C; Denaturation, 5 min at 95 °C; Incubation, 175 min at 60 °C; Incubation at 20 °C until use). The complete cycle should take approximately 5 h. Converted DNA can be left in the thermal cycler overnight without any loss of performance. Since the bisulfite reaction is not overlaid with mineral oil, only thermal cyclers with heated lids are suitable for this procedure. It is important to use PCR tubes that close tightly.

3.5 Cleanup of Bisulfite Converted DNA

1. Due to the sensitivity of nucleic acid amplification technologies, take the following precautions when handling MinElute DNA spin columns to avoid cross-contamination between samples.

 – Pipet carefully the sample or solution into the MinElute DNA spin column without wetting the rim of the column.

 – Avoid touching the MinElute DNA spin column membrane with the pipet tip.

 – Always change pipet tips between liquid transfers. We recommend the use of aerosol-barrier pipet tips.

 – Open one MinElute DNA spin column at a time, and take care to avoid generating aerosols.

- Wear gloves throughout the entire procedure. In case of contact between gloves and sample, change gloves immediately.

2. All centrifugation steps should be carried out at room temperature (15–25 °C).

3. Centrifuge the PCR tubes containing the bisulfite reactions.

4. Transfer the complete bisulfite reactions to clean 1.5 ml microcentrifuge tubes. Keep the PCR tubes for washing and use one pipet tip per PCR tube for transferring bisulfite reactions and washing solutions. Transfer of precipitates in the solution will not affect the performance or yield of the reaction.

5. Add 310 μl freshly prepared Buffer BL containing 10 μg/ml carrier RNA to each PCR tube.

6. Transfer to the corresponding 1.5 ml tube containing the bisulfite converted DNA and mix the solutions by vortexing and then centrifuge briefly.

7. Add 250 μl of ethanol (100 %) to each PCR tube.

8. Transfer to the corresponding 1.5 ml tube containing the bisulfite converted DNA.

9. Mix the solutions by pulse vortexing for 15 s.

10. Centrifuge briefly to remove the drops from inside the lid.

11. Place the necessary number of MinElute DNA spin columns and collection tubes in a suitable rack.

12. Transfer the entire mixture from each tube into the corresponding MinElute DNA spin column. Keep the 1.5 ml tube for washing.

13. Centrifuge the spin columns at maximum speed for 1 min.

14. Discard the flow-through and place the spin columns back into the collection tubes.

15. Add 500 μl Buffer BW (wash buffer) to each 1.5 ml tubes.

16. Transfer to the corresponding spin column.

17. Centrifuge at maximum speed for 1 min.

18. Discard the flow-through, and place the spin columns back into the collection tubes.

19. Add 500 μl Buffer BD (desulfonation buffer) to each spin column, and incubate for 15 min at room temperature (15–25 °C). If there are precipitates in Buffer BD, avoid transferring them to the spin columns (*see* **Notes 2** and **3**).

20. Centrifuge the spin columns at maximum speed for 1 min.

21. Discard the flow-through and place the spin columns back into the collection tubes.

22. Add 500 μl Buffer BW to each spin column and centrifuge at maximum speed for 1 min.

23. Discard the flow-through and place the spin columns back into the collection tubes.

24. Repeat **step 21** once.

25. Add 250 μl of ethanol (96–100 %) to each spin column and centrifuge at maximum speed for 1 min.

26. Place the spin columns into new 2 ml collection tubes and centrifuge the spin columns at maximum speed for 1 min to remove any residual liquid.

27. Place the spin columns into clean 0.2 ml microcentrifuge tubes and place them in 1.5 ml microcentrifuge tubes without covers.

28. Add 15 μl Buffer EB (elution buffer) directly onto the center of each spin-column membrane and close the lids gently.

29. Incubate the spin columns at room temperature for 1 min.

30. Centrifuge for 1 min at $15,000 \times g$ to elute the DNA (*see* **Notes 4** and **5**).

Proceed to PCR amplification directly in the tube containing the 15 μl of purified DNA.

3.6 General Considerations About PCR Amplification of the Bisulfite Converted DNA

This protocol enables PCR amplification from single MII oocytes or single 2-cell embryos DNA with an efficiency of up to 80 %. Several points are critical for optimization of the PCR amplification step. The first one is the size of the DNA fragment to be amplified. Even though the use of a DNA protect buffer provided with the EpiTect Plus LyseAll Bisulfite Kit limits efficiently the fragmentation usually associated with bisulfite treatment of DNA at high temperatures and low pH values, best results were obtained with DNA sequences not exceeding 500 bp in length. Two round of PCR are required. The use of a hot start Taq polymerase is strictly necessary. PCR primer design for converted sequences of interest (for example http://www.dsi.univ-paris5.fr/bio2/PCRProg.html) is also critical. Ideally, the primers should amplify as little as 10 pg of bisulfite modified DNA (if no signal or very weak signal is obtained, a two-bases shift in the primer sequence to the -5′ or -3′ end is sometimes sufficient to get a nice signal); the melting temperature (MT) should be close to 60 °C ± 2 °C; the size may vary from 20 to 35 mers. The GC % is not critical, as far as the melting temperatures range the recommended values. The forward and the reverse primers of a pair may have varied size, but we recommend that they exhibit close melting temperatures. The second PCR may be primed efficiently with nested or semi nested primers as well.

Table 1
Primer table

Gene	Primers	Size	MT (°C)
KCNQ1OT1 DMR EF	5'-TGTTTTTGTAGTTTATATGGAAGGGTTAA-3'	29	60.5
KCNQ1OT1 DMR ER	5'-CTCACCCCTAAAAACTTAAAACCTC-3'	25	59.8
KCNQ1OT1 DMR IF	5'-GTTAGGGAAGTTTTAGGGTGTGAAT-3'	25	60.0
KCNQ1OT1 DMR IR	5'-AAACATACCAAACCACCCACCTAACAAA-3'	28	66.9
H19 DMR EF	5'-TTYGTAGGGTTTTTGGTAGGTATAGAGTT-3'	29	60.9/62.6
H19 DMR ER	5'-ATAAATATCCTATTCCCAAATAACCCC-3'	27	60.8
H19 DMRIF	5'-AGTATATGGGTATTTTTGGAGGTTTTT-3'	27	59.7

EF external forward primer, *ER* external reverse primer, *IF* internal forward primer, *IR* internal reverse primer
KCNQ1OT1 DMR: GenbankU90095, 66,536–66800 bp; *H19* DMR: Genbank: AF125183, 7,875–8,096 bp
Oct4: Genbank AJ297527, –78/+131; *Nanog*: Ensembl gene ID ENSG00000111704, –724/–435

Table 2
Primer table

Oct4 EF	5'-ATTTTAGTTTGGGTAATAAAGTGAGATTTTG-3'	31	61.4
Oct4 ER	5'-CCCACACCTCAAAACCTAACC-3'	21	60.6
Oct4 IF	5'-GAGGGAGAGAGGGGTTGAGTAG-3'	22	60.6
Oct4 IR	5'-CCTCCAAAAAAACCTTAAAAACTTAAC-3'	27	60.1
Nanog EF	5'-ATTATAATTTTTGTTTTTTAGGTTTAAGGG-3'	30	59.1
Nanog ER	5'-ACAACAAACCTAAAAACAAACCCA-3'	25	61.9
Nanog IF	5'-TTTGTTTTTTAGGTTTAAGGGATTTTTT-3'	28	60.8
Nanog IR	5'-CCAACTTTTAAATCAAAAATATAATTCAAC-3'	30	59.2

EF external forward primer, *ER* external reverse primer, *IF* internal forward primer, *IR* internal reverse primer
KCNQ1OT1 DMR: GenbankU90095, 66,536–66,800 bp; *H19* DMR: Genbank: AF125183, 7,875–8,096 bp
Oct4: Genbank AJ297527, –78/+131; *Nanog*: Ensembl gene ID ENSG00000111704, –724/–435

Examples of PCR primers that are highly efficient in priming PCR amplification are given in Tables 1 and 2.

1. The whole 15 µl of purified bisulfite converted DNA obtained are utilized for PCR1, in a final volume of 50 µl. The nested PCR is run on 1 µl of PCR1 amplified product. PCR1 and nested PCR are usually run for 40 cycles.

2. The PCR products are submitted to electrophoresis on 1.5 % agarose gel/ethidium bromide to fractionate DNA fragments.

3.7 Extraction and Purification of Amplified Bisulfite Converted DNA

Extraction and purification of amplified bisulfite converted DNA was done using E.Z.N.A. Gel Extraction Kit from OMEGA bio-tek. The key to this system is the HiBind® matrix that specifically, but reversibly, binds DNA or RNA under optimized conditions allowing proteins and other contaminants to be removed. Nucleic acids are easily eluted with deionized water or a low salt buffer. DNA fragments between 70 bp and 20 kb can be recovered with yields exceeding 85 %.

1. Following electrophoresis, when adequate separation of bands has occurred, carefully excise the DNA fragment of interest using a wide, clean, sharp scalpel. Minimize the size of the gel slice by removing extra agarose. Limit exposure of your PCR product to shortwave UV light to avoid formation of pyrimidine dimers. Use a glass plate between the gel and UV source. If possible, only visualize the PCR product with a long-wave UV source.

2. Determine the appropriate volume of the gel slice by weighing it in a clean 1.5 ml microcentrifuge tube. Assuming a density of 1 g/ml, the volume of gel is derived as follows: a gel slice of mass 0.3 g will have a volume of 0.3 ml.

3. Add 1 volume Binding Buffer (XP2).

4. Incubate at 60 °C for 7 min or until the gel has completely melted. Vortex or shake the tube every 2–3 min.

5. Insert a HiBind® DNA Mini Column in a 2 ml Collection Tube.

6. Add no more than 700 μl DNA/agarose solution from **step 5** to the HiBind® DNA Mini Column.

7. Centrifuge at $10,000 \times g$ for 1 min at room temperature.

8. Discard the filtrate and reuse collection tube.

9. Repeat **steps 7–9** until the entire sample has been transferred to the column.

10. Add 300 μl Binding Buffer (XP2).

11. Centrifuge at maximum speed ($\geq 13,000 \times g$) for 1 min at room temperature.

12. Discard the filtrate and reuse collection tube.

13. Add 700 μl SPW Wash Buffer. SPW Wash Buffer must be diluted with 100 % ethanol prior to use.

14. Centrifuge at maximum speed for 1 min at room temperature.

15. Discard the filtrate.

16. Repeat **steps 13–15** for a second SPW Wash Buffer wash step. Perform the second wash step for any salt sensitive downstream applications.

17. Centrifuge the empty HiBind® DNA Mini Column for 2 min at maximum speed to dry the column matrix. It is important to dry the HiBind® DNA Mini Column matrix before elution. Residual ethanol may interfere with downstream applications.

18. Transfer the HiBind® DNA Mini Column to a clean 1.5 ml microcentrifuge tube.

19. Add 100 μl Elution Buffer or deionized water directly to the center of the column membrane. The efficiency of eluting DNA from the HiBind® DNA Mini Column is dependent on pH. If eluting DNA with deionized water, make sure that the pH is around 8.5.

20. Let sit at room temperature for 2 min.

21. Centrifuge at maximum speed for 1 min. This represents approximately 70 % of bound DNA. An optional second elution will yield any residual DNA, though at a lower concentration.

22. Store DNA at –20 °C. The OD is measured on 80 μl.

3.8 Ligation of the Amplified Bisulfite Converted DNA to pGEM-T Easy Vector

1. Optimize Insert:Vector Molar Ratio. To calculate the appropriate amount of PCR product (insert) to include in the ligation reaction, use the following equation:

$$\frac{\text{ng of vector} \times \text{kb size of insert}}{\text{kb size of vector}} \times \text{insert : vector molar ratio}$$
$$= \text{ng of insert}$$

Example of insert:vector ratio calculation:
How much 0.5 kb PCR product should be added to a ligation in which 50 ng of 3.0 kb vector will be used if a 3:1 insert:vector molar ratio is desired?

$$\frac{50\text{ng vector} \times 0.5\text{kb insert}}{3.0\text{kb vector}} \times \frac{3}{1} = 25\text{ng insert}$$

Using the same parameters for a 1:1 insert:vector molar ratio, 8.3 ng of a 0.5 kb insert would be required.

2. Briefly centrifuge the pGEM®-T vector and Control Insert DNA tubes to collect the contents at the bottom of the tubes (*see* **Note 6**).

3. Set up ligation reactions as described below in 0.5 ml tubes known to have low DNA-binding capacity (*see* **Note 7**).

Reaction component	Standard reaction	Positive control	Background control
2× Rapid Ligation Buffer, T4 DNA Ligase	5 μl	5 μl	5 μl
pGEM®-T Vector (50 ng)	1 μl	1 μl	1 μl
PCR product	3 μl	–	–
Control Insert DNA	–	2 μl	–
T4 DNA Ligase (3 Weiss units/μl)	1 μl	1 μl	1 μl
Nuclease-free water to a final volume of	10 μl	10 μl	10 μl

4. Vortex the 2× Rapid Ligation Buffer vigorously before each use (*see* **Note 8**).

5. Mix the reactions by pipetting.

6. Incubate the reactions overnight at 4 °C (*see* **Note 9**).

3.9 Transformation

1. Important considerations for successful T-Vector cloning. The ligation of fragments with a single-base overhang can be inefficient, so it is essential to use cells with a transformation efficiency of at least 1×10^8 cfu/μg DNA in order to obtain a reasonable number of colonies.

2. Use JM109 High Efficiency Competent Cells ($\geq 1 \times 10^8$ cfu/μg DNA) for transformations in order to obtain a reasonable number of colonies. Other host strains may be used, but they should be compatible with blue/white color screening and standard ampicillin selection (*see* **Note 10**).

3. Prepare two LB/ampicillin/IPTG/X-Gal plates for each ligation reaction, plus two plates for determining transformation efficiency. Equilibrate the plates to room temperature.

4. Centrifuge the tubes containing the ligation reactions to collect the contents at the bottom.

5. Add 2 μl of each ligation reaction to a sterile (17×100 mm) polypropylene tube or a 1.5 ml microcentrifuge tube on ice.

6. Set up another tube on ice with 0.1 ng uncut plasmid for determination of the transformation efficiency of the competent cells. We have found that use of larger (17×100 mm) polypropylene tubes increases transformation efficiency.

7. Remove tube(s) of frozen JM109 High Efficiency Competent Cells from storage and place in an ice bath until just thawed (about 5 min).

8. Mix the cells by gently flicking the tube. Avoid excessive pipetting, as the competent cells are extremely fragile.

9. *Carefully* transfer 50 μl of cells into each tube prepared in **step 2** (use 100 μl of cells for determination of transformation efficiency).

10. *Gently* flick the tubes to mix and place them on ice for 20 min.

11. Heat-shock the cells for 45–50 s in a water bath at exactly 42 °C (*do not shake*).

12. Immediately return the tubes to ice for 2 min.

13. Add 950 μl room-temperature SOC medium to the tubes containing cells transformed with ligation reactions and 900 μl to the tube containing cells transformed with uncut plasmid (LB broth may be substituted, but colony number may be lower).

14. Incubate for 1.5 h at 37 °C with shaking (~150 rpm).

15. Plate 100 μl of each transformation culture onto duplicate LB/ampicillin/IPTG/X-Gal plates. For the transformation control, a 1:10 dilution with SOC medium is recommended for plating. If a higher number of colonies is desired, the cells may be pelleted by centrifugation at $1,000 \times g$ for 10 min, resuspended in 200 μl of SOC medium, and 100 μl plated on each of two plates.

16. Incubate the plates overnight (16–24 h) at 37 °C. If 100 μl is plated, approximately 100 colonies per plate are routinely seen using competent cells that are 1×10^8 cfu/μg DNA. Longer incubations or storage of plates at 4 °C (after 37 °C overnight incubation) may be used to facilitate blue color development. White colonies generally contain inserts; however, inserts may also be present in blue colonies (*see* **Note 11**).

3.10 Blue/White Selection of Recombinants and Screening Transformants for Inserts and Sequence Analysis

The pGEM®-T vectors are high-copy-number vectors containing T7 and SP6 RNA polymerase promoters flanking a multiple cloning region within the α-peptide coding region of the enzyme β-galactosidase. Inactivation of the α-peptide by insertion allows identification of recombinants by blue/white screening on indicator plates. Successful cloning of an insert into the pGEM®-T Vector interrupts the coding sequence of β-galactosidase (*see* **Notes 12** and **13**).

1. Identify recombinant clones by color screening on indicator plates Pick up white colonies and plate them in a new plate with numbered marked squares.

2. For each colony, drop the tip that served to pick up the colony in a 0.2 ml tube containing 50 μl of PCR mix with primers complementary to T7 and SP6 promoters.

3. Perform an amplification round with an annealing temperature of 55 °C, for 40 cycles.

4. Subject 15 μl of the PCR product to electrophoresis on 1.5 % agarose gel/ethidium bromide to control the size of the insert.

5. Use directly 20 μl for Sanger sequencing.

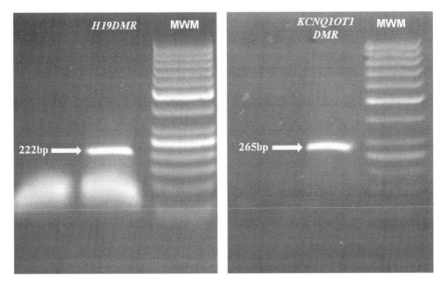

Fig. 1 Analysis of the methylation of *H19* DMR and *KCNQ1OT1* DMR at the single human oocyte level. Amplification of bisulfite converted DNA from a single MII oocyte. The primers are as specified above. *MWM* molecular weight marker

H19DMR	KCNQ1OT1 DMR
□■□□□□□□□□□□□□□□□□	■□■■■■■■■■■■■■■■■■■
□□□□□□□□□□□□□□□■□■	■■■□■■■■■■■■■■■■■■■□□

Fig. 2 Bisulfite sequencing analysis of *H19* DMR or *KCNQ1OT1* DMR in single oocytes. Each line represents a single allele. A *black square* indicates a methylated CpG, and an *open square* denotes an unmethylated CpG. Ten clones were sequenced for each PCR product. Identical sequences are counted once as they likely represent the same initial DNA strand

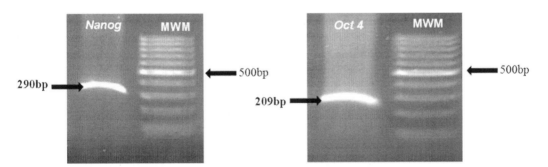

Fig. 3 Analysis of the methylation of *Oct4* and *Nanog* promoters at the 2-cell human embryo level. Amplification of bisulfite converted DNA from a single 2-cell human ICSI embryo. Following duplex PCR, sequences of *Oct 4* and *Nanog* promoters were amplified by their specific pairs of internal primers. The primers are as specified above. *MWM* molecular weight marker

3.11 Results Obtained Utilizing This Procedure

1. Analysis of the methylation of *H19 DMR* and *KCNQ1OT1 DMR* at the single human oocyte level. *See* Figs. 1 and 2.

2. Analysis of the methylation of Oct4 and Nanog promoters at the 2-cell human embryo level. *See* Figs. 3 and 4.

Nanog, 2-cell embryo n°104	Oct 4, 2-cell embryo n°104
□□□□□□□□	■■■■■■■□■■
□□□□□□□□	■■■■■■■■□■
	□□■■■■■■□■
	■■■■■■■□■□■

Fig. 4 Bisulfite sequencing analysis of *Nanog* and *Oct4* promoters in a single 2-cell human ICSI embryo. Each line represents a single allele. A *black square* indicates a methylated CpG, an *open square* denotes an unmethylated CpG. Ten clones were sequenced for each PCR product. Identical sequences are counted once as they likely represent the same initial DNA strand

4   Notes

1. If necessary, heat the Bisulfite Mix–RNase-free water solution to 60 °C and vortex again. Do not place dissolved Bisulfite Mix on ice.

2. The bottle containing Buffer BD should be closed immediately after use to avoid acidification from carbon dioxide in air.

3. It is important to close the lids of the spin columns before incubation.

4. As little as 10 μl Buffer EB can be used for elution if a higher DNA concentration is required, but the yield will be reduced by approximately 20 %. Do not elute with less than 10 μl Buffer EB as the spin column membrane will not be sufficiently hydrated.

5. If the purified DNA is to be stored for up to 24 h, we recommend storage at 2–8 °C. For storage longer than 24 h, we recommend storage at –20 °C.

6. T-Overhangs for Easy PCR Cloning. This step was performed with the pGEM-T Easy Vector System from Promega, according to the manufacturer recommendations. The pGEM®-T vectors are linearized vectors with a single 3′-terminal thymidine at both ends. The T-overhangs at the insertion site greatly improve the efficiency of ligation of PCR products by preventing recircularization of the vector and providing a compatible overhang for PCR products generated by certain thermostable polymerases.

7. Use only the T4 DNA Ligase supplied with this system to perform pGEM®-T vector ligations. Other commercial preparations of T4 DNA ligase may contain exonuclease activities that may remove the terminal deoxythymidines from the vector.

8. 2× Rapid Ligation Buffer contains ATP, which degrades during temperature fluctuations. Avoid multiple freeze–thaw cycles and exposure to frequent temperature changes by making single-use aliquots of the buffer.

9. The pGEM®-T vector System is supplied with 2× Rapid Ligation Buffer. Ligation reactions using this buffer may be

incubated for 1 h at room temperature. The incubation period may be extended to increase the number of colonies after transformation. Generally, an overnight incubation at 4 °C produces the maximum number of transformants.

10. Use of super high-efficiency competent cells (e.g., XL10 Gold® Ultracompetent Cells) may result in a higher background of blue colonies

11. Colonies containing β-galactosidase activity may grow poorly relative to cells lacking this activity. After overnight growth, the blue colonies may be smaller than the white colonies, which are approximately 1 mm in diameter. Blue color will become darker after the plate has been stored overnight at 4 °C.

12. The characteristics of the PCR products cloned into the vectors can significantly affect the ratio of blue/white colonies obtained. Usually clones containing PCR products produce white colonies, but blue colonies can result from PCR fragments that are cloned in-frame with the lacZ gene. Such fragments are usually a multiple of three base pairs long (including the 3′-A overhangs) and do not contain in-frame stop codons. There have been reports of DNA fragments up to 2 kb that have been cloned in-frame and have produced blue colonies. Even if your PCR product is not a multiple of three bases long, the amplification process can introduce mutations (deletions or point mutations) that may result in blue colonies.

13. *See* Fig. 5.

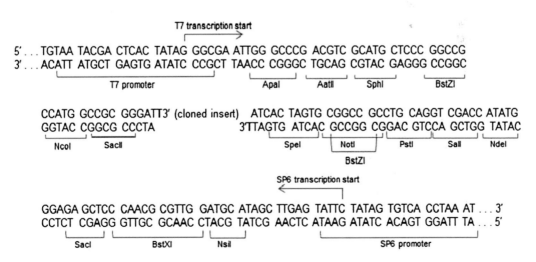

Fig. 5 The promoter and multiple cloning sequence of the pGEM®-T Vector. The *top strand* corresponds to the RNA synthesized by T7 RNA polymerase. The *bottom strand* corresponds to the RNA synthesized by SP6 RNA polymerase. The pGEM®-T vectors contain numerous restriction sites within the multiple cloning regions. The pGEM®-T vector cloning region is flanked by recognition sites for the enzyme BstZI. Alternatively, a double-digestion may be used to release the insert from either vector

References

1. Sanford JP, Clark HJ, Chapman VM et al (1987) Differences in DNA methylation during oogenesis and spermatogenesis and their persistence during early embryogenesis in the mouse. Genes Dev 1:1039–1046

2. Clark SJ, Harrison J, Paul CL, Frommer M (1994) High sensitivity mapping of methylated cytosines. Nucleic Acids Res 15:2990–2997

3. Al-Khtib M, Perret A, Ibala-Romdhane S, Khoueiry R, Blachère T, Lornage J, Lefèvre A (2011) Vitrification at the GV stage, followed by IVM, does not affect the methylation profile of H19 and KCNQ1OT1 imprinting centres in human oocytes. Fertil Steril 95: 1955–1960

4. Ibala-Romdhane S, Al-Khtib M, Khoueiry R, Blachère T, Guérin JF, Lefèvre A (2011) Analysis of H19 methylation in control and abnormal embryos sperm and oocytes. Eur J Hum Genet 11:1138–1143

5. Al Khtib M, Blachère T, Guérin JF, Lefèvre A (2012) Methylation profile of the promoters of Nanog and POU5F1 in ICSI human embryos. Hum Reprod 10:2948–2954

Chapter 17

Micro Chromatin Immunoprecipitation (µChIP) from Early Mammalian Embryos

John Arne Dahl and Arne Klungland

Abstract

Chromatin immunoprecipitation (ChIP) is a powerful method for mapping protein–DNA interactions in vivo. Genomic localization of histone modifications, transcription factors, and other regulatory proteins can be revealed by ChIP. However, conventional ChIP protocols require the use of large numbers of cells, which prevents the application of ChIP to rare cell types. We have developed ChIP assays suited for the immunoprecipitation of histone proteins or transcription factors from small cell numbers. Here we describe a rapid, yet sensitive micro (µ)ChIP protocol producing high signal to noise ratio output, suitable for as few as 100 cells. This chapter provides a detailed protocol for µChIP from early mammalian embryos, also suitable for any sample of limited numbers of cells. Minor modifications of this optimized high signal to noise ChIP protocol make it a reliable tool for the use with any cell number ($100-10^7$).

Key words Chromatin immunoprecipitation, ChIP, Histone, Acetylation, Methylation, Epigenetics, Embryo, Inner cell mass, Trophectoderm

1 Introduction

Throughout the embryo developmental program, different cells and tissues acquire diverse functions and morphologies. These differences result from various programs of gene expression and not changes in the DNA sequence. Post translational modifications of histones is one of the main epigenetic mechanisms involved in regulating proper gene expression during mammalian development and works in concurrence with highly regulated binding of transcription factors and other regulatory proteins to *cis*-regulatory sequences such as promoters and enhancers. The accessibility of *cis*-regulatory sequences is dependent on chromatin structure where the organization of modified histone proteins and DNA play a key role.

From the one-cell to the blastocyst stage, there are changes in global DNA methylation and histone modifications [1–12]. At the morula stage, the first lineage allocation event in mammalian

Nathalie Beaujean et al. (eds.), *Nuclear Reprogramming: Methods and Protocols*, Methods in Molecular Biology, vol. 1222, DOI 10.1007/978-1-4939-1594-1_17, © Springer Science+Business Media New York 2015

embryogenesis occurs, and by the blastocyst stage the first two lineages have been set up: the inner cell mass (ICM, giving rise to embryonic lineages) and the trophectoderm (TE, giving rise to extraembryonic lineages) [2, 13]. We are just starting to understand how cell fate, leading to two lineages, is determined [2, 8, 13–20].

Insight on the significance of interactions between proteins and DNA in the context of gene regulation and cell differentiation in health and disease has immensely been increased because of the chromatin immunoprecipitation (ChIP) strategy that was established in the late 1980s [21, 22]. The ChIP technique allows for a protein of interest to be selectively immunoprecipitated from a cell lysate to identify the DNA sequences associated with it in vivo. ChIP has been extensively used to reveal the localization of transcription factors, post-translationally modified histones, histone variants or chromatin modifying enzymes on a given locus or genome-wide [23–36].

Although ChIP is a powerful technique, it has for a long time remained a lengthy and laborious procedure requiring large cell numbers, making it incompatible with analysis of early embryos. Several ChIP assays have recently been developed to overcome the requirement for large cell numbers and have allowed for the study of histone modifications and transcription factors, within a handful of loci or at the genome scale, from cell numbers in the range of 100 to 100,000 [37–46]. We have developed improved ChIP assays to simplify and shorten the procedure and make ChIP feasible to locus specific or genome-wide analysis of limited numbers of cells. Our µChIP methodology has been applied to explore histone modifications in the ICM and TE of the early mammalian embryo [14, 47]. This chapter provides detailed instructions for µChIP from ICM and TE, including a description of the isolation of ICM and TE from bovine blastocysts. Nonetheless, the described µChIP procedure is suitable for other mammalian early embryo samples, and likely for any sample of limited number of cells. The notes section describes details on modifications of this optimized high signal to noise ChIP protocol to make it a reliable tool for the use with any cell number ($100–10^7$).

We divide the µChIP method into six major procedures: (1) cross-linking of DNA and proteins, (2) binding of antibodies to paramagnetic beads, (3) chromatin preparation, (4) immunoprecipitation and washes, (5) DNA isolation and purification, and (6) quantitative PCR and analysis of data.

2 Materials

2.1 Materials

1. Filter pipette tips (10, 200, 1,000 µL).
2. Siliconized pipette tips.

3. 200-μL PCR tubes in 8-tube strip format.

4. 0.6 and 1.5-mL centrifuge tubes.

5. Magnetic rack for 200-μL PCR tube strips.

6. Magnetic holder for 1.5-mL tubes.

7. Microblades to bisect the blastocysts.

8. Sonicator fitted with 2-mm diameter probe (*see* **Note 1**).

9. Tube rotator for "head-over-tail" motion, placed at 4 °C.

10. Refrigerated tabletop centrifuge with a swing-bucket rotor for 1.5-mL tubes.

11. Adapters for 0.6-mL tubes to be inserted into 1.5-mL tube rotor.

12. Minicentrifuge with rotors for 1.5-mL tubes and PCR tubes.

13. Vortex with a PCR tube holder.

14. Thermomixer.

15. Heating block.

16. Thermal cycler with real time capacity for quantitative PCR.

17. Stereomicroscope when dissecting blastocysts.

2.2 Reagents

1. 36.5 % formaldehyde.

2. Dynabeads® Protein A and/or G (Invitrogen) (*see* **Note 2**).

3. 5 M NaCl.

4. 400 mM EGTA.

5. 500 mM EDTA.

6. 1 M Tris–HCl, pH 7.5.

7. 1 M Tris–HCl, pH 8.0.

8. Triton X-100.

9. Na-deoxycholate.

10. 1.25 M Glycine stock solution in PBS.

11. Acrylamide carrier.

12. 20 mg/mL Proteinase K solution in Milli-Q water.

13. Protease inhibitor mix (Sigma-Aldrich).

14. 100 mM PMSF stock solution in 100 % ethanol.

15. For anti-acetylated epitope ChIPs: 1 M sodium butyrate stock solution in Milli-Q water.

16. Trinitrobenzenesulfonic acid (TNBS).

17. Polyvinyl alcohol (PVA).

18. Guinea pig complement serum.

19. Phosphate buffered saline (PBS).

20. Sodium dodecyl sulfate (SDS).

21. Phenol–chloroform–isoamylalcohol (25:24:1).

22. Chloroform–isoamylalcohol (24:1).

23. 3 M NaAc.

24. 96 % ethanol at −20 °C.

25. 70 % ethanol at −20 °C.

26. Reagents for real time quantitative PCR (*see* **Note 3**).

27. Deionized water.

28. Crushed ice.

29. Antibodies of choice. Use ChIP-grade antibodies when available.

30. Anti-dinitrophenyl-BSA.

2.3 Buffers

1. Lysis buffer: 50 mM Tris–HCl, pH 8.0, 10 mM EDTA, 1 % (wt/vol) SDS, protease inhibitor mix (1:100 dilution from stock), 1 mM PMSF, 20 mM Na-butyrate. Protease inhibitor mix, PMSF, and Na-butyrate are added immediately before use.

2. RIPA buffer: 10 mM Tris–HCl, pH 7.5, 140 mM NaCl, 1 mM EDTA, 0.5 mM EGTA, 1 % (vol/vol) Triton X-100, 0.15 % (wt/vol) SDS, 0.1 % (wt/vol) Na-deoxycholate.

3. RIPA ChIP buffer: 10 mM Tris–HCl, pH 7.5, 140 mM NaCl, 1 mM EDTA, 0.5 mM EGTA, 1 % (vol/vol) Triton X-100, 0.15 % (wt/vol) SDS, 0.1 % (wt/vol) Na-deoxycholate, protease inhibitor mix (1:100 dilution from stock), 1 mM PMSF, 20 mM Na-butyrate. Protease inhibitor mix, PMSF, and Na-butyrate are added immediately before use.

4. TE buffer: 10 mM Tris–HCl, pH 8.0, 1 mM EDTA.

5. Elution buffer: 20 mM Tris–HCl, pH 7.5, 5 mM EDTA, 50 mM NaCl, 20 mM Na-butyrate.

6. Complete elution buffer: 20 mM Tris–HCl, pH 7.5, 5 mM EDTA, 50 mM NaCl, 20 mM Na-butyrate, 1 % (wt/vol) SDS, 140 μg/mL proteinase K. SDS and proteinase K are added just before use.

3 Methods

3.1 Isolation of ICM and TE

1. Select blastocysts with a clearly visible ICM for isolation of ICM and TE cells (e.g., day 8 after in vitro fertilization for bovine blastocysts). Work with groups of for example 10–30 blastocysts for each round of isolation (*see* **Note 4**).

2. Wash blastocysts in drops of PBS containing 1 mg/mL polyvinyl alcohol (PVA).

3. Manually bisect blastocysts under a stereomicroscope by lowering down a microblade while the ICM is oriented in the

apical part (*see* **Note 5**). This results in a pure TE cell fraction and a second fraction of ICM cells with some adhering TE cells.

4. Immediately wash the TE cell fractions twice in PBS with PVA, collect them in a 0.6 mL tube in 500 μL PBS, and continue with cross-linking of DNA and proteins as described under Subheading 3.2 (*see* **Note 6**).

5. Remove adhering TE cells from the ICM by the following immunosurgery procedure: Incubate ICMs in trinitrobenzenesulfonic acid for 8–10 min at 4 °C, wash twice in PBS with 1 mg/mL PVA, incubate in anti-dinitrophenyl-BSA for 20 min at room temperature, and culture in guinea pig complement serum under oil in the incubator to lyse the remaining TE cells (30 min at 39 °C for bovine blastocysts).

6. Wash ICMs twice in PBS containing 1 mg/mL PVA to remove TE cells and collect ICMs in a 0.6 mL tube in 500 μL PBS and continue directly with cross-linking of DNA and proteins as described under Subheading 3.2.

3.2 Cross-Linking of DNA and Proteins

1. Add 13.5 μL of 36.5 % formaldehyde to each of the 0.6 mL tubes containing either TEs or ICMs in 500 μL PBS for a final concentration of 1 % (vol/vol), mix by very gentle vortexing, and incubate for 8 min at room temperature (*see* **Note 7**). Repeat the gentle vortexing twice during the incubation step (*see* **Note 8**) (*see* **Note 9**).

2. Add 57 μL of the 1.25 M glycine stock (final concentration 125 mM), mix by gentle vortexing, and incubate for 5 min at room temperature. Repeat vortexing once during the incubation step.

3. Pellet the ICMs or TEs in each tube by centrifugation at $550 \times g$ for 10 min at 4 °C in a swing-bucket rotor with soft deceleration settings (*see* **Note 10**).

4. Carefully aspirate and discard the supernatant. Let ~30 μL of the solution remain with the cell pellet to ensure that there is no loss of embryo cells.

5. Resuspend ICM or TE cells in 500 μL ice-cold PBS by gentle vortexing and centrifuge as above.

6. Repeat the washing procedure (**Step 4** and **5**). After the last wash, carefully remove more supernatant than for the washing steps and leave 10–20 μL of PBS in the bottom of the tube to avoid disturbing the pellet (*see* **Note 11**).

7. Snap freeze in liquid nitrogen and store at –80 °C (*see* **Note 12**). Collect as many pools of embryo cells as required for the downstream ChIP experiments.

3.3 Binding of Antibodies to Paramagnetic Beads

To save time, binding of antibodies to paramagnetic beads is carried out during chromatin preparation (Subheading 3.4) and can if necessary be prolonged until all chromatin samples are ready for immunoprecipitation. If more convenient, one may also start it the day before the immunoprecipitation.

1. Vortex the stock of paramagnetic Dynabeads® thoroughly for 2 min to ensure the stock bead suspension is homogenous before pipetting (*see* **Notes 13 and 14**).

2. For 18 ChIPs (9 for ICM and 9 for TE), including two negative controls (one for each cell type), pipette 150 µL of well-suspended Dynabeads® stock solution into a 1.5-mL tube, place the tube in a suited magnetic rack and allow beads to be captured (*see* **Note 15**).

3. Discard the buffer, remove the tube from the magnet and add 500 µL of RIPA buffer.

4. Vortex to a homogenous suspension, quickly spin in a minicentrifuge to recover any liquid caught in the lid, capture the beads, remove the supernatant and add another 500 µL of RIPA buffer.

5. Vortex, capture the beads, remove the supernatant and add 140 µL RIPA buffer.

6. Vortex the beads to a homogenous suspension and place the tube on ice.

7. Aliquot 90 µL of RIPA buffer into 200-µL PCR tubes (use one tube per ChIP or negative control) and place on ice. We recommend using 200-µL PCR tubes in an 8-tube strip format. These fits in the magnetic rack.

8. Vortex thoroughly the washed bead-slurry, add 5 µL of bead suspension to each of the 200 µL PCR tubes and add 2 µg of a well-characterized antibody to each tube (*see* **Notes 16–18**). To the negative controls, do not add any antibody, or add 2 µg of a pre-immune antibody preferably of the same isotype as the ChIP antibodies (*see* **Note 19**).

9. Incubate at 40 rpm on a "head-over-tail" tube rotator for 2 h at 4 °C.

3.4 Chromatin Preparation

Here we describe the procedure to prepare chromatin from the pooling of 6 batches of frozen cross-linked embryo cells. It is however also suited for single tubes as well as for the pooling of fewer or more batches of cells.

1. Collect a desired number of cross-linked and frozen pools of embryo cells from –80 °C and place on dry-ice in an insulated box. Specific details outlined below are for chromatin preparation from the pooling of 6 batches of frozen cross-linked

embryo cells (*see* **Note 20**). Work with one cell type at a time for the lysis step, e.g., first TE, then ICM. (For adjusting or scaling up cell lysis to other conditions *see* **Note 21**).

2. Prepare a total volume of 160 μL lysis buffer for each pool of TE or ICM.

3. Move one tube containing frozen cross-linked embryo cells from dry-ice to wet-ice, leave for ~5 s to allow any frozen drops of liquid in the lid or on the tube wall to initiate melting, briefly spin in a minicentrifuge to recover any frozen drops of liquid in the bottom of the tube, immediately place the tube back on ice.

4. Add 60 μL of room temperature lysis buffer, let the frozen liquid melt, and then mix well with gentle pipetting using a siliconized pipette tip (*see* **Note 22**).

5. Let the pipette tip (tip only) stay in the tube containing lysed embryo cells while bringing the next tube of embryo cells from dry-ice to wet-ice and follow the procedure described at **step 3**.

6. Transfer the lysis buffer and cell lysate from the first tube to the second tube using the siliconized pipette tip that was stored in the first tube, let frozen liquid melt and mix well by gentle pipetting. Again keep the pipette tip in the tube. Immediately close the lid of the first tube to limit evaporation.

7. Successively wash through the remaining four tubes to collect all cellular material in the last tube of the series.

8. Add 34 μL of fresh lysis buffer to the first tube of the series using a clean pipette tip, switch to the previously used siliconized pipette tip and successively wash through all six tubes.

9. Add 66 μL of additional lysis buffer, successively wash through all tubes as described above and collect all cell lysate in the last tube of the series, keeping the same siliconized tip throughout the procedure.

10. Measure the final volume by comparing to identical tubes containing known volumes of lysis buffer. This volume will allow for the calculation of the final SDS concentration in the lysate which will be the basis for the calculation of how much SDS to have in the RIPA ChIP buffer that will be added before the immunoprecipitation step (*see* **Note 23** for details).

11. Prepare a sufficient volume of RIPA ChIP buffer with an SDS concentration that will result in a 0.23 % final concentration of SDS in the ChIP reaction (*see* **Note 24**). For ten reactions (eight ChIPs, one negative control and one input control) prepare 1.5 mL of RIPA ChIP buffer containing 0.15 % SDS. From this preparation, 1,305 μL will be added stepwise to the 245 μL lysate after sonication (**step 12**), resulting in a final volume of 1.55 mL ChIP-ready chromatin (*see* **Note 23** for details on the lysate volume and SDS concentration).

12. With the probe sonicator (*see* **Note 1**), use the following sonication procedure: Sonicate cell lysate on ice for 3×30 s, with 30 s pauses on ice between each 30 s session and with pulse settings at 0.5 s cycles and 30 % power. Repeat for each tube of cell lysate while leaving the sonicated samples on ice (*see* **Note 25**).

13. Add 275 μL RIPA ChIP buffer from **step 11** to the tube which contains ~245 μL sonicated lysate and mix by vortexing.

14. Centrifuge at $12,000 \times g$ for 10 min at 4 °C, carefully aspirate the supernatant (chromatin) and transfer it into a clean 1.5 mL tube chilled on ice, leaving ~50 μL of supernatant with the (invisible) pellet (*see* **Note 26**).

15. Add 500 μL RIPA ChIP buffer to the remaining pellet, mix by vortexing, and centrifuge at $12,000 \times g$ for 10 min at 4 °C.

16. Collect the supernatant, leaving ~10 μL with the pellet and pool it with the first supernatant.

17. Add 150 μL RIPA ChIP buffer to the remaining 10 μL, mix by vortexing, and perform another 3×30 s sonication of the pellet as described at **step 12**.

18. Add 390 μL RIPA ChIP buffer from **step 11**, mix by vortexing, centrifuge at $12,000 \times g$ for 10 min at 4 °C, carefully aspirate the supernatant, leaving ~10 μL with the pellet and pool it with the two previous supernatants. Mix by vortexing, briefly spin on a minicentrifuge to remove any liquid caught in the tube lid and place tube with 1.55 mL ChIP-ready chromatin on ice (*see* **Note 27**).

19. Remove the 200-μL PCR tubes containing paramagnetic beads coated with antibodies from the rotator (*see* **step 9** in Subheading 3.3).

20. Quickly spin in a minicentrifuge to recover any liquid caught in the lid, place the tubes in a magnetic rack on ice to capture the beads, remove the supernatant, and add 100 μL of RIPA buffer.

21. Gently shake PCR tubes by hand to bring the beads in suspension, incubate at 40 rpm on a "head-over-tail" rotator for 4 min at 4 °C, spin in a minicentrifuge to recover any liquid caught in the lid, place the tubes in a magnetic rack on ice to capture beads, and remove the supernatant.

22. Immediately aliquot 150 μL of ChIP-ready chromatin from **step 18** into each tube, e.g., eight chilled 200-μL PCR tubes (in strip format) and one tube for the negative control.

23. Add 150 μL of chromatin to a 1.5-mL tube chilled on ice to be used as input control. Store at 4 °C until **step 4** in Subheading 3.6.

3.5 Immunoprecipitation and Washes

1. Remove the PCR tube strip from the magnetic rack to allow the antibody-bead complexes to be released into the chromatin solution, gently shake PCR tubes by hand to fully bring the beads in suspension, and place the tubes on a rotator at 40 rpm for 3 h at 4 °C (*see* **Note 28**).

2. Briefly centrifuge the tubes in a minicentrifuge to bring down any solution trapped in the lid, and capture the immune complexes by placing the PCR tubes in a magnetic rack cooled on ice.

3. Discard the supernatant containing the unbound fraction, add 100 μL ice-cold RIPA buffer and remove the tubes from the magnetic rack to allow beads with captured immune complexes to be released into the buffer. Resuspend the beads with immune complexes by gentle manual shaking and rotate PCR tubes at 40 rpm for 4 min at 4 °C.

4. Repeat **steps 2** and **3** twice.

5. Briefly spin the tubes in a minicentrifuge to collect any liquid trapped in the lid before placing the PCR tubes in the magnetic rack.

6. Remove the supernatant, add 100 μL TE buffer, resuspend the beads with immune complexes by moderate manual agitation and incubate on a rotator at 4 °C for 4 min at 40 rpm.

7. Centrifuge the PCR tubes in a minicentrifuge for 1 s.

8. Place the PCR tubes on ice (do not use the magnetic rack), resuspend to a homogenous suspension with a pipette, and transfer the content of each tube into separate clean 200-μL PCR tubes on ice (*see* **Note 29**).

9. Add 50 μL TE buffer to each of the emptied tubes and mix with a pipette to get any remaining beads in suspension, transfer the remaining beads to the same tube as the previous 100 μL of bead suspension (*see* **Note 30**).

10. Capture the complexes by placing the PCR tubes in the magnetic rack and remove the TE buffer. Immediately continue with Subheading 3.6 (*see* **Note 31**).

3.6 DNA Isolation and Purification

DNA elution from immune complexes, cross-link reversal and protein digestion are combined to a single step (*see* **Note 32**).

1. Add 150 μL complete elution buffer to each of the PCR tubes from Subheading 3.5, **step 10**.

2. Vortex tube strips, using a suitable holder for PCR tubes, to make sure that all beads are well suspended.

3. Incubate for 2 h and 30 min on a Thermomixer at 68 °C, 1,300 rpm (*see* **Note 33**).

4. Prepare the input sample while ChIP samples are incubating:, Add 200 μL of elution buffer and 1 μL of the proteinase K

solution to input chromatin samples, vortex well, and incubate for 2 h and 30 min on a heating block at 68 °C.

5. Remove PCR tubes from the Thermomixer and centrifuge with a minicentrifuge.

6. Use the magnetic rack to capture the beads, collect the supernatant containing ChIP DNA and place it into a clean 1.5-mL tube.

7. Add 150 μL complete elution buffer to the PCR tubes to wash the beads and incubate on the Thermomixer for 5 min at 68 °C, 1,300 rpm.

8. Remove PCR tubes from the Thermomixer, use the magnetic rack to capture the beads, collect the supernatant and pool it with the first supernatant.

9. Add 190 μL elution buffer to the eluted ChIP material, resulting in a final volume of 490 μL for all ChIP samples and negative controls.

10. Remove input samples from the heating block, briefly centrifuge and transfer each input sample to a clean 1.5 mL tube. Wash the incubated tube with 190 μL elution buffer and combine with the input sample transferred to the clean tube, resulting in a final volume of 490 μL for input controls (*see* **Note 34**).

11. For all ChIP samples, negative controls and input controls: Extract DNA once using an equal volume of phenol–chloroform–isoamylalcohol (25:24:1) (490 μL), vortex and shake all tubes vigorously, centrifuge at $15,000 \times g$ for 5 min at 20 °C to separate the aqueous and organic phases and transfer 460 μL of the aqueous (top) phase to a clean 1.5-mL tube (*see* **Note 35**).

12. Extract DNA with a 460 μL volume of chloroform–isoamylalcohol, centrifuge at $15,000 \times g$ for 5 min at 20 °C, and collect 400 μL of the aqueous (top) phase and add to a clean1.5-mL tube (*see* **Note 36**).

13. Add 44 μL of 3 M NaAc (pH 7.0), 11 μL of 0.25 % (wt/vol) acrylamide carrier, and 1 mL 100 % ethanol chilled at −20 °C. Mix well and place at −80 °C for at least 1 h (*see* **Note 37**).

14. Thaw the tubes and immediately centrifuge at $20,000 \times g$ for 15 min at 4 °C.

15. Visually inspect each tube to ensure that there is a visible, white leaf-like pellet in the bottom of the tube.

16. Remove the supernatant while leaving ~50 μL with the pellet, add 1 mL of 70 % ethanol prechilled at −20 °C and vortex briefly to get the DNA pellet to detach from the bottom of the tube. Centrifuge tubes at $20,000 \times g$ for 10 min at 4 °C. Repeat this washing step once more.

17. Remove as much as possible of the supernatant without touching the pellet and dry the pellet to get rid of all ethanol (*see* **Note 38**).

18. Dissolve the DNA in a desired volume of a suitable solvent. For example use 50–100 μL of TE buffer when doing 500 cells per ChIP and applying qPCR for downstream analysis, and let dissolve on ice over night (*see* **Note 39**). If not using the DNA for further analysis right away, store at −80 °C (*see* **Note 40**).

3.7 Quantitative PCR and Analysis of Data

1. Prepare 25 μL qPCR reactions (6.5 μL Milli-Q water; 12.5 μL SYBR Green Master Mix (2×); 0.5 μL forward primer (20 μM stock); 0.5 μL reverse primer (20 μM stock); 5 μL DNA template) for all ChIP samples, inputs and negative controls, with each primer pair.

2. Prepare a dilution series to make a standard curve with genomic DNA. For early embryo samples one can use genomic DNA sampled from a suitable organ from individuals of the same genetic lineage. Include a relevant range of DNA concentrations (e.g., 0.005–20 ng/μL) to cover the concentration range of your ChIP DNA samples. Use 5 μL of standard curve DNA in each PCR. Make a standard curve for each primer pair for every PCR plate.

3. Prepare a real time PCR program with 40 cycles on your real time PCR system.

4. Acquire amplification data with your real time PCR software.

5. Calculate the starting amount of DNA in each ChIP, input and control sample using the standard curve.

6. Export the data into appropriate spreadsheets for calculations.

7. Calculate the amount of precipitated ChIP DNA relative to input DNA as ([Amount of ChIP DNA]/[Amount of input DNA]) × 100. Use the mean of duplicate qPCRs for calculations of the mean of independent ChIPs and express the data as percentage precipitated DNA relative to input DNA (*see* **Note 41**).

4 Notes

1. We recommend using the Sartorius Labsonic M sonicator, or similar fitted with a 2-mm diameter probe. To the best of our knowledge the Labsonic M sonicator is no longer available for purchase from Sartorius, but Hielscher is the maker of this sonicator and have it available under the product name UP100H.

2. Use Dynabeads® Protein A beads with rabbit IgGs and Dynabeads® Protein G with mouse IgGs for highest IgG binding capacity. These Dynabeads® give very low levels of unspecific

binding compared to for example sepharose and agarose beads. The beads should be well suspended before pipetting.

3. The sensitivity and efficiency of ready to use reagents for real time quantitative PCR varies from supplier to supplier and also between different kits from the same supplier. Make sure to test real time quantitative PCR reagents using a relevant DNA template before assessing valuable ChIP samples from embryos. We have obtained good results with for example IQ SYBR® Green (Bio-Rad).

4. The size of the group of blastocysts one may work with depends on the investigators practice and skill. It is important to work quickly to make sure that embryo cells will not be affected more than necessary by the handling performed prior to cross-linking of DNA and proteins. All handling outside of the in vivo situation involve a potential to affect the biology of the cells.

5. Scratches in the petri dish were made to avoid slipping of the embryos during bisection.

6. It might be helpful to be two persons working together at this step to ensure immediate progress of the TE samples through the protocol and cross-linking of the DNA and proteins, while performing the ICMs immunosurgery without waiting.

7. Formaldehyde cross-links proteins to proteins and proteins to DNA located within 2 Å of each other [48]. To simplify the cross-linking step and enhance cell recovery, we cross-link cells in suspension unless cell harvest of adherent cells is feasible due to high cell numbers. The optimal time of cross-linking might vary with the protein to be immunoprecipitated, but for most applications, 8–10 min cross-linking should be sufficient.

8. Up to 500,000 cells can be cross-linked using the described procedure. If scaling up further use a concentration of 1–2 million cells per 1 mL. When obtaining cell pellets of 10 μL or more one may remove all of the supernatant before snap freezing in liquid nitrogen. For cell numbers of 500,000–2 million cells use a 1.5 mL eppendorf tube. If scaling up cell numbers further we use a 50 mL tube for cross-linking and transfer appropriate aliquots of cross-linked cells to eppendorf tubes after glycine treatment and PBS washes.

9. If assessing acetylated epitopes, one should consider using a histone deacetylase inhibitor: Add 20 μL of 1 M Na-butyrate per 1 mL of cells in PBS immediately before cross-linking.

10. Use a swing-out rotor to avoid a smear of cells on the tube wall. The swing-out rotor ensures that cells are pelleted in the very bottom of the tube and importantly aid to reduce cell loss.

11. Take great care to avoid cell loss during washes. Start out with a 200-μL pipette tip and change to a smaller and smaller tip when approaching the bottom of the tube, ending up with a 10 μL pipette tip. Remove liquid from the top, just touching the surface only. Keep in mind that the pellet is not visible with the naked eye.

12. Pellets of cross-linked cells can be stored at −80 °C for at least 1 year, however the procedure works equally well for frozen and freshly prepared cross-linked cells.

13. It is important to ensure the stock bead suspension is homogenous before pipetting as one otherwise will not be able to reproducibly add the same amount of beads per ChIP reaction in subsequent experiments. Additionally one will affect the concentration of paramagnetic beads in the stock suspension.

14. Use protein A coated beads with rabbit IgGs and protein G coated beads with mouse IgGs.

15. Make sure to start out with an excess of beads to avoid running short on beads when distributing beads for all the ChIP reactions. The volumes for a certain number of ChIPs provided in this setup is just meant to be an example and one can scale up or down as suitable for any setup one would like to carry out.

16. Preferably purchase ChIP grade antibodies. Further we recommend to ask the commercial provider for batch specific information on antibody performance as some companies provide the same data set (obtained with one antibody batch) for validation of specificity for every antibody batch covered by the same catalogue number. Because each batch of polyclonal antibodies has been produced in a different animal one should not simply assume every batch to perform equally well.

17. Vortex thoroughly the washed bead-slurry before starting pipetting to ensure homogenous suspension, then make sure to vortex gently before every transfer of 5 μL of beads as the beads sediment rather quickly.

18. When scaling up to 10,000 cells or more per ChIP one may increase bead volume to 10 μL and antibody amount to 2.4 μg. For 1–10 million cells one may further double (or more) these numbers to increase the amount of precipitated ChIP DNA.

19. Comparisons in our lab have shown that the background precipitation is equal for negative control samples of either bead-only or of beads coated with pre-immune antibodies when working with Dynabeads.

20. Here we describe chromatin preparation from the pooling of six batches of frozen cross-linked embryo cells. Use as many pools of embryo cells as required for the downstream ChIP experiments. The number of embryos required depends on the

number of ChIPs one will set up and the number of loci one would like to assess per ChIP. We recommend to allow minimum a 100-cell-equivalent of chromatin per ChIP to assess no more than three loci by duplicate qPCR. Increasing the cell number to for example 500 cells per ChIP would ensure less stochastic variation and allow for more loci to be investigated.

21. Adjusting or scaling up cell lysis to other conditions is possible. If one is starting out with 500–50,000 cells, having all cross-linked cells in the same tube, use the following procedure for cell lysis: Add 120 μL room temperature lysis buffer, vortex for 2×5 s, leave on ice for 5 min, and vortex again for 2×5 s. Ensure that no liquid is trapped in the lid. Continue with sonication. The following procedure applies for cell numbers above 50,000 cells: Add the larger of 120 μL or a 5x volume of lysis buffer as compared to the cell pellet size, and mix well with gentle pipetting and sonicate a maximum of 250 μL in each tube. The number of sonication sessions as described at **step 12** in Subheading 3.4 needs to be optimized for each cell type and cell concentration. If starting out with less than 500 cells see our previously published description on chromatin preparation from 100 cells [42].

22. Take care to avoid foaming as foaming can affect recovery of cell lysate and later on also influence chromatin fragmentation efficiency.

23. For example the final volume may be ~245 μL indicating that ~14 μL was present in each tube containing cross-linked and frozen pools of embryo cells (6×14 μL = 84 μL; 84 μL + 160 μL lysis buffer = 244 μL). A final volume of 245 μL results in an SDS concentration of 0.65 %.

24. Depending on the antibody, epitope, cell type, and chromatin concentration one may find that the optimal final SDS concentration for ChIP is between 0.1 and 0.3 %. We recommend optimizing stringency for each antibody. However, in our experience 0.15–0.25 % is often optimal. For optimization, start out with a less stringent situation as a reference for titrations of increasing stringency. One may also try to titrate the NaCl concentration if increasing the SDS concentration is not beneficial and results in reduced precipitation rather than an increase in the signal to noise ratio.

25. Sonication of cell lysates was done to produce ~400–500 bp chromatin fragments. Fragment size of DNA was assessed by quantitative (q) PCR, as previously described [43]. The sonication regime must be optimized for each cell type, in particular for primary cells. Do not allow cell lysate to foam as foaming reduces sonication efficiency. To avoid foaming, ensure that the sonicator probe is placed deep enough in the tube, a few millimeters from the bottom of the tube, or try to reduce sonication intensity.

26. To avoid aspirating the sedimented material, which is not visible with the naked eye, leave ~50 µL supernatant in the tube after aspiration.

27. Whenever possible we prefer to use the chromatin fresh, however chromatin can be stored at −80 °C for several months.

28. This step can be carried out overnight at 4 °C if necessary, but prolonged incubation may enhance background. If scaling up to 10,000 cells or more per ChIP and using 10 µL of beads in an incubation volume of 100 µL, it is sufficient to incubate for 2 h on a rotator.

29. Take care to avoid scraping of the pipette tip on the tube wall. Try to avoid contact between the pipette tip and the tube wall, as the tube wall is a source of unspecific binding of chromatin. ChIP signal to noise ratio is increased if one can exclude unspecifically bound chromatin from the sample.

30. The additional washing of the ChIP tubes with 50 µL of TE buffer may be omitted when doing ChIP from 50,000 cells or more.

31. Never allow beads to dry out as this could affect the outcome of your immunoprecipitation.

32. In addition to save quite a bit of time, this design results in less sample-to-sample variation and increased DNA recovery as compared to the conventional strategy keeping the three steps of DNA elution, cross-link reversal and protein digestion separated.

33. We recommend using a suitable adhesive tape to secure the 200-µL PCR tube strips to the Thermomixer plate as the vigorous shacking at 1,300 rpm otherwise may result in tubes coming loose and in some cases in loss of sample.

34. The transfer of the input sample to a clean tube after incubation at 68 °C reduces the chances of leakage during phenol–chloroform–isoamylalcohol extraction. The softening of the plastic resulting from heating of a tube makes it more likely to experience leakage as compared to when using a clean non-heated tube.

35. It is helpful to first use a 1-mL pipette to recover the top 260 µL and then switch to a 200-µL pipette to recover the next 200 µL. Take great care and very slowly aspirate the aqueous phase to avoid getting any of the interphase or organic phase.

36. Use filtered tips when adding phenol–chloroform–isoamylalcohol (25:24:1) and chloroform–isoamylalcohol to prevent dripping during transfer.

37. DNA can be stored at −80 °C for several days if more convenient. We routinely keep our samples at −80 °C until all samples have become solid to ensure high reproducibility.

38. Inspect tubes and pellets every 5 min when drying and add TE or other solvent as soon as the pellet is completely dry. Time required for drying is dependent on the volume of ethanol left with the pellet, but one can expect a range from 15 min to 1 h.

39. TE volume for dissolving DNA depends on the number of cells in the ChIP. Note that low DNA concentrations may lead to degradation of the DNA more rapidly than what would be expected for higher concentrations. Thus, we recommend immediately using DNA for PCR or other downstream applications for ChIPs from 1,000 cells or less. If this is not feasible we recommend storing the ChIP-DNA at –80 °C. Avoid EDTA containing buffers if applying a whole genome strategy for downstream analysis. For high throughput sequencing we recommend using 11 μL of a 1 mM Tris buffer. If one would like to save time one may dissolve DNA on a Thermomixer for 1 h at 30 °C, 800 rpm. However, for precious samples to reduce the risk of sample degradation we recommend dissolving DNA on ice.

40. We have previously described a quick Chelex-100-mediated DNA purification suited for small cell numbers [42]. This procedure can, with minor modifications, be used for any cell numbers and is a fast way to purify DNA for downstream qPCR analysis. However, we recommend purifying ChIP DNA with the phenol–chloroform–isoamylalcohol extraction method when considering genome-wide analysis. Keep in mind that the Chelex-100 mediated DNA purification results in melting of double stranded DNA due to boiling of samples.

41. If no PCR signal is detected, several factors might be implicated. (1) There is too little chromatin in the ChIP assay. Solution: increase the amount of cells or chromatin (note that it can be difficult to extract all chromatin from certain primary cell types); (2) the ChIP did not work. Solution: use ChIP-grade antibodies if available; do an antibody titration; include a positive control antibody; (3) the PCR did not work. Solution: set up control qPCRs with the same primers for a dilution series including relevant concentrations from genomic DNA and optimize PCR conditions by assessing the resulting standard curve. If ChIP-PCR signals are weaker than expected, there might not be sufficient DNA template. If variations in PCR signal intensity are observed between ChIP replicates, this may be due to: (1) inconsistent chromatin preparations between samples. Solution: make sure that insoluble debris is removed by sedimentation after chromatin fragmentation; avoid debris when aspirating the chromatin supernatant; (2) inconsistent sonication. Solution: practice sonication using larger cell numbers (e.g., 100,000–10 million cells) until

fragmentation is reproducible; (3) inconsistent amounts of Dynabeads between samples: ensure magnetic beads are well suspended, to homogeneity, while pipetting; (4) not enough and variable amounts of ChIP-DNA template resulting in increased stochastic variation (may be observed at high Ct values). Solution: increase the amount of ChIP-DNA template in the PCR and ensure consistency between replicates; make sure that ethanol-precipitated DNA is fully dissolved before setting up PCR.

Acknowledgements

Our work is supported by the Norwegian Cancer Society. We are thankful to Dr. Adam Robertson for reading and commenting on important parts of this chapter.

References

1. Arney KL, Erhardt S, Drewell RA, Surani MA (2001) Epigenetic reprogramming of the genome—from the germ line to the embryo and back again. Int J Dev Biol 45(3):533–540

2. Morgan HD, Santos F, Green K, Dean W, Reik W (2005) Epigenetic reprogramming in mammals. Hum Mol Genet 14 Spec No 1:R47–58

3. van der Heijden GW et al (2005) Asymmetry in histone H3 variants and lysine methylation between paternal and maternal chromatin of the early mouse zygote. Mech Dev 122(9): 1008–1022

4. Torres-Padilla ME, Bannister AJ, Hurd PJ, Kouzarides T, Zernicka-Goetz M (2006) Dynamic distribution of the replacement histone variant H3.3 in the mouse oocyte and preimplantation embryos. Int J Dev Biol 50(5):455–461

5. Torres-Padilla ME, Parfitt DE, Kouzarides T, Zernicka-Goetz M (2007) Histone arginine methylation regulates pluripotency in the early mouse embryo. Nature 445(7124):214–218

6. Mayer W, Niveleau A, Walter J, Fundele R, Haaf T (2000) Demethylation of the zygotic paternal genome. Nature 403(6769):501–502

7. Oswald J et al (2000) Active demethylation of the paternal genome in the mouse zygote. Curr Biol 10(8):475–478

8. Santos F, Hendrich B, Reik W, Dean W (2002) Dynamic reprogramming of DNA methylation in the early mouse embryo. Dev Biol 241(1): 172–182

9. Santos F, Peters AH, Otte AP, Reik W, Dean W (2005) Dynamic chromatin modifications characterise the first cell cycle in mouse embryos. Dev Biol 280(1):225–236

10. Santos F et al (2003) Epigenetic marking correlates with developmental potential in cloned bovine preimplantation embryos. Curr Biol 13(13):1116–1121

11. Iqbal K, Jin SG, Pfeifer GP, Szabo PE (2011) Reprogramming of the paternal genome upon fertilization involves genome-wide oxidation of 5-methylcytosine. Proc Natl Acad Sci U S A 108(9):3642–3647

12. Wossidlo M et al (2011) 5-Hydroxymethyl-cytosine in the mammalian zygote is linked with epigenetic reprogramming. Nat Commun 2:241

13. Rossant J, Tam PP (2009) Blastocyst lineage formation, early embryonic asymmetries and axis patterning in the mouse. Development 136(5):701–713

14. Dahl JA, Reiner AH, Klungland A, Wakayama T, Collas P (2010) Histone H3 lysine 27 methylation asymmetry on developmentally-regulated promoters distinguish the first two lineages in mouse preimplantation embryos. PLoS One 5(2):e9150

15. Smith ZD et al (2012) A unique regulatory phase of DNA methylation in the early mammalian embryo. Nature 484:339–344

16. Smith ZD, Meissner A (2013) DNA methylation: roles in mammalian development. Nat Rev Genet 14(3):204–220

17. Niwa H et al (2005) Interaction between Oct3/4 and Cdx2 determines trophectoderm differentiation. Cell 123(5):917–929

18. Surani MA, Hayashi K, Hajkova P (2007) Genetic and epigenetic regulators of pluripotency. Cell 128(4):747–762

19. Dean W et al (2001) Conservation of methylation reprogramming in mammalian development: aberrant reprogramming in cloned embryos. Proc Natl Acad Sci U S A 98(24): 13734–13738

20. Erhardt S et al (2003) Consequences of the depletion of zygotic and embryonic enhancer of zeste 2 during preimplantation mouse development. Development 130(18):4235–4248

21. Solomon MJ, Larsen PL, Varshavsky A (1988) Mapping protein-DNA interactions in vivo with formaldehyde: evidence that histone H4 is retained on a highly transcribed gene. Cell 53: 937–947

22. Hebbes TR, Thorne AW, Crane-Robinson C (1988) A direct link between core histone acetylation and transcriptionally active chromatin. EMBO J 7(5):1395–1402

23. Boyer LA, Lee TI, Cole MF, Johnstone SE, Levine SS (2005) Core transcriptional regulatory circuitry in human embryonic stem cells. Cell 122:947–956

24. O'Neill LP, Turner BM (1995) Histone H4 acetylation distinguishes coding regions of the human genome from heterochromatin in a differentiation-dependent but transcription-independent manner. EMBO J 14(16): 3946–3957

25. Bernstein BE, Kamal M, Lindblad-Toh K, Bekiranov S (2005) Genomic maps and comparative analysis of histone modifications in human and mouse. Cell 120:169–181

26. Azuara V, Perry P, Sauer S, Spivakov M (2006) Chromatin signatures of pluripotent cell lines. Nat Cell Biol 8:532–538

27. Loh YH et al (2006) The Oct4 and Nanog transcription network regulates pluripotency in mouse embryonic stem cells. Nat Genet 38: 431–440

28. Lee TI, Jenner RG, Boyer LA, Guenther MG, Levine SS (2006) Control of developmental regulators by Polycomb in human embryonic stem cells. Cell 125:301–313

29. Guenther MG, Levine SS, Boyer LA, Jaenisch R (2007) A chromatin landmark and transcription initiation at most promoters in human cells. Cell 130:77–88

30. Mikkelsen TS et al (2007) Genome-wide maps of chromatin state in pluripotent and lineage-committed cells. Nature 448(7153):553–560

31. Zhao XD et al (2007) Whole-genome mapping of histone H3 Lys4 and 27 trimethylations reveals distinct genomic compartments in human embryonic stem cells. Cell Stem Cell 1:286–298

32. Dixon JR et al (2012) Topological domains in mammalian genomes identified by analysis of chromatin interactions. Nature 485(7398): 376–380

33. Hunkapiller J et al (2012) Polycomb-like 3 promotes polycomb repressive complex 2 binding to CpG islands and embryonic stem cell self-renewal. PLoS Genet 8(3):e1002576

34. Shen Y et al (2012) A map of the *cis*-regulatory sequences in the mouse genome. Nature 488(7409):116–120

35. Hon GC et al (2013) Epigenetic memory at embryonic enhancers identified in DNA methylation maps from adult mouse tissues. Nat Genet 45(10):1198–1206

36. Gamble MJ, Frizzell KM, Yang C, Krishnakumar R, Kraus WL (2010) The histone variant macroH2A1 marks repressed autosomal chromatin, but protects a subset of its target genes from silencing. Genes Dev 24(1):21–32

37. O'Neill LP, VerMilyea MD, Turner BM (2006) Epigenetic characterization of the early embryo with a chromatin immunoprecipitation protocol applicable to small cell populations. Nat Genet 38(7):835–841

38. Dahl JA, Collas P (2007) Q2ChIP, a quick and quantitative chromatin immunoprecipitation assay, unravels epigenetic dynamics of developmentally regulated genes in human carcinoma cells. Stem Cells 25(4):1037–1046

39. Attema JL et al (2007) Epigenetic characterization of hematopoietic stem cell differentiation using miniChIP and bisulfite sequencing analysis. Proc Natl Acad Sci U S A 104(30): 12371–12376

40. Acevedo LG et al (2007) Genome-scale ChIP-chip analysis using 10,000 human cells. Biotechniques 43(6):791–797

41. Dahl JA, Collas P (2008) MicroChIP—a rapid micro chromatin immunoprecipitation assay for small cell samples and biopsies. Nucleic Acids Res 36(3):e15

42. Dahl JA, Collas P (2008) A rapid micro chromatin immunoprecipitation assay (microChIP). Nat Protoc 3(6):1032–1045

43. Dahl JA, Reiner AH, Collas P (2009) Fast genomic muChIP-chip from 1,000 cells. Genome Biol 10(2):R13

44. Adli M, Zhu J, Bernstein BE (2010) Genome-wide chromatin maps derived from limited numbers of hematopoietic progenitors. Nat Methods 7(8):615–618

45. Goren A et al (2010) Chromatin profiling by directly sequencing small quantities of

immunoprecipitated DNA. Nat Methods 7(1):
47–49

46. Gilfillan GD et al (2012) Limitations and possibilities of low cell number ChIP-seq. BMC Genomics 13:645

47. Herrmann D, Dahl JA, Lucas-Hahn A, Collas P, Niemann H (2013) Histone modifications and mRNA expression in the inner cell mass and trophectoderm of bovine blastocysts. Epigenetics 8(3):281–289

48. Orlando V (2000) Mapping chromosomal proteins in vivo by formaldehyde-crosslinked-chromatin immunoprecipitation. Trends Biochem Sci 25(3):99–104

Chapter 18

Assessing Reprogramming by Chimera Formation and Tetraploid Complementation

Xin Li, Bao-long Xia, Wei Li, and Qi Zhou

Abstract

Pluripotent stem cells can be evaluated by pluripotent markers expression, embryoid body aggregation, teratoma formation, chimera contribution and even more, tetraploid complementation. Whether iPS cells in general are functionally equivalent to normal ESCs is difficult to establish. Here, we present the detailed procedure for chimera formation and tetraploid complementation, the most stringent criterion, to assessing pluripotency.

Key words Pluripotent stem cell, Chimera, Tetraploid complementation, Reprogramming, Embryonic development

1 Introduction

Chimeric mice were first created in the 1960s by Kristoph Tarkowski and Beatrice Mintz [1], through aggregating two eight-cell embryos, and were then produced by Richard Gardner and Ralph Brinster by injection of cells into blastocysts [2]. This revolutionary new technique showed a new way for introducing any kind of cells into the host embryo, and defining the in vivo development properties of these cells

Authentic stem cells such as Embryonic Stem cells (ES) or Induced Pluripotent Stem cells (iPS) have three cardinal properties: unlimited symmetrical self-renewal, comprehensive ability to contribute to chimeras, and generation of germline transmission gametes [3]. However, 4n complementation is regarded as the most solid criterion for ES cell pluripotency. The same criterion is also accepted for induced pluripotent stem cells reprogrammed from somatic cells [4]. Here we describe the different steps required to prepare the cells to be tested and to prepare the blastocysts,

---

Xin Li, Bao-long Xia, and Wei Li contributed equally to this work.

Nathalie Beaujean et al. (eds.), *Nuclear Reprogramming: Methods and Protocols,* Methods in Molecular Biology, vol. 1222, DOI 10.1007/978-1-4939-1594-1_18, © Springer Science+Business Media New York 2015

either diploid for chimeras or tetraploid for 4n complementation. We will also explain the procedure of cells injection into blastocysts; blastocysts that can then be transferred back into recipient mothers to evaluate their developmental potential.

2 Materials

2.1 Material for Cell Preparation

1. Gelatin-coated dishes: dissolve 0.1 g gelatin powder in PBS by autoclaving for 15 min, then store at 4 °C. To prepare the gelatin-coated dish, add 5 ml 0.1 % gelatin solution to the 100 mm dish and store at 37 °C for 3–4 days (*see* **Note 1**).

2. 1× Phosphate-buffered saline (PBS) Ca^{2+} and Mg^{2+} free.

3. 0.25 % liquid trypsin (1×) with EDTA·4Na.

4. DMEM supplemented with 10 % Fetal Bovine Serum (FBS).

5. 100 mm culture dishes and 10 ml disposable pipettes.

6. Low speed centrifuge (for 15 ml centrifuge tube).

7. Cell culture medium (either for ES or iPS cells) (*see* **Note 2**).

8. 1.5 ml microtubes.

2.2 Mice

All experiments requiring animal handling should be conducted in accordance with international animal protection guidelines and local animal protection laws.

1. Pregnant mice, 8–10 weeks old, either on day 1.5 (for two-cell collection) or on day 3.5 (for blastocyst collection) after mating.

2. Recipient pseudopregnant mice, 8–10 weeks old, 2.5 days post-coitum.

2.3 Material for Embryo Collection and Transfer

1. Disposable consumables: pads or absorbent towels, forceps and scissors, 1 ml syringes, 27G and 30G blunt-end needles, 60 and 35 mm culture dishes, embryo handling pipettes, sutures or wound clips.

2. Equipment: dissecting microscope, electrofusion generator, incubator at 37 °C with CO_2 supply.

3. Commercially available embryo media: M2 medium, M16 medium, KSOM.

4. CZB medium: 81.62 mM NaCl, 4.83 mM KCl, 1.7 mM $CaCl_2 \cdot 2H_2O$, 1.18 mM KH_2PO_4, 1.18 mM $MgSO_4 \cdot 7H_2O$, 25.12 mM $NaHCO_3$, 31.30 mM sodium lactate, 0.27 mM sodium pyruvate, 0.11 mM EDTA (disodium salt), 1 mM Glutamine, 100 mM sodium penicillin, 0.7 mM streptomycin, 5 mM BSA in water. Store at 4 °C for up to 2 weeks.

5. Hepes-CZB medium (H-CZB): same as in **step 8** except that 0.48 mM Hepes should be added and that the $NaHCO_3$ concentration should be reduced to 5 mM.

6. Others reagents: ethanol, glucose, mineral oil.

2.4 Tools for Blastocyst Injection

1. Micromanipulation equipment: Piezo, inverted microscope with phase-contrast optics or differential interference equipped with two micromanipulator control units (including one with oil, *see* **Note 3** and Fig. 1) and a cooled stage (*see* **Note 4**).

2. Micromanipulation chamber (*see* **Note 5**).

3. Glass holding pipettes and injection pipettes (*see* **Note 6**). The injection pipette should have a flat tip with a 15 μm internal diameter. The holding pipettes have a flat tip with a 30 μm internal diameter.

3 Methods

3.1 Preparation of the Cells

Cells used for injection should be growing exponentially and at a sub-confluent density. They should be prepared 30–60 min before injection into blastocysts (*see* **Note 1**).

1. Remove the cell culture medium and wash the cells once with pre-warmed PBS (at 37 °C).

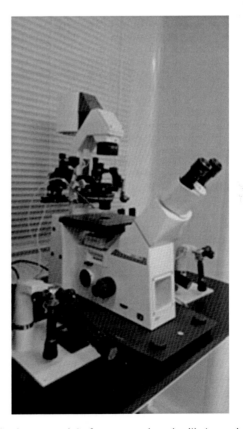

Fig. 1 Inverted microscope interference equipped with two micromanipulator control units, a Piezo and a cooled stage

2. Add 1 ml of 0.25 % trypsin, incubate the cells for 3–4 min at 37 °C.

3. Stop the trypsin reaction by adding 4 ml of DMEM supplemented with 10 % FBS.

4. Resuspend the cells by pipetting them gently up and down.

5. Place the suspension in a 100 mm culture dish and incubate the cells at 37 °C for 15–20 min (*see* **Note 8**).

6. Carefully collect all the medium containing the cells in suspension and transfer to 15 ml centrifuge tube (*see* **Note 9**).

7. Centrifuge the cell suspension at 200 × *g* for 3–5 min.

8. Discard the supernatant and resuspend the cells thoroughly in 300 µl of cell culture medium.

9. Transfer into a 1.5 ml microtube and place the tube on ice. The cells are now ready for injection (*see* **Note 10**).

3.2 Preparing Blastocysts for Micromanipulation

3.2.1 Collection of Diploid Blastocysts

1. Euthanize the mice (pregnant day 3.5 female mice) and place them on their back on a pad or absorbent towel (*see* **Note 11**).

2. Spray abdomen with ethanol.

3. Pick up the skin with the blunt forceps and cut through the lower abdominal wall and internal fascia. Then cut along the lateral walls, to expose the gut.

4. Displace the abdominal cavity contents, by moving it aside over the thorax.

5. Expose the cervix, uterine horns, oviducts, and ovaries.

6. Pull the uterotubal junction toward the side of the mouse and remove the fatty mesometrium adhering to the oviduct and uterus.

7. Put the uterus and oviduct in a 60 mm culture dish with M2 medium.

8. Fill a 1 ml syringe with M2 medium and attach the blunt-end 30G needle.

9. Hold the uterus with forceps (at the oviduct end), and insert the needle at the uterotubal junction and flush the uterus.

10. Collect the embryos and wash in M2 medium with an embryo handling pipette, under the dissecting microscope.

11. Culture the embryos at 37 °C under 5 % CO_2 in drops of KSOM or M16 medium covered with mineral oil in a 35 mm dish.

3.2.2 Preparing Blastocysts for Tetraploid Complementation

1. Euthanize the mice on day 1.5 after mating for two-cell embryos collection (*see* **Note 12**).

2. Gently collect the oviducts with fine scissors and forceps from the abdominal cavities as described from **steps 2–6** in Subheading 3.2.1.

3. Immediately wash twice with H-CZB medium and put into a new 60 mm dish with H-CZB medium.

4. Flush out the two-cell stage embryos from the oviducts with a blunt-end 30G needle, a 1 ml syringe, and H-CZB medium.

5. Collect the two-cell embryos and place them in pre-warmed CZB medium for short culture.

6. Electrofusion is then performed with an electrofusion generator using the following parameters: DC 5 V, 10 s; AC 50 V, 35 μs for two cycles; in order to electrofuse the blastomeres of the two-cell embryos.

7. Collect the pulsed embryos in CZB medium with an embryo handling pipette under the dissecting microscope, cover the medium with mineral oil, and culture at 37 °C under 5 % CO_2.

8. One hour later, remove the embryos that did not fuse.

9. Transfer the tetraploid embryos into CZB medium supplemented with 5.56 mM glucose.

10. Culture for another 48 h at 37 °C under 5 % CO_2, until the blastocyst stage.

11. Microinjection of "pluripotent" cells to be tested can then be performed.

3.3 Blastocyst Injection

1. Place the micromanipulation chamber filled with embryo manipulation medium under mineral oil on the microscope stage (see Note 13).

2. Use an embryo handling pipette to add hundreds of cells into the chamber.

3. Transfer the expanded blastocysts into the chamber (see Note 14).

4. Pick up 10–15 cells into the injection pipette and position them near the tip, in a minimal amount of medium.

5. Hold a single blastocyst with the holding pipette and move it toward the center of the microscope field.

6. Position the blastocyst with the inner cell mass at 9 o'clock position, near the holding pipette (see Fig. 2).

7. Put the tip of the injection needle on the surface of the blastocyst (see Fig. 3).

8. Push the injection pipette through the zona pellucida and the trophectoderm layer with a brief pulse of Piezo to insert the injection needle into the cavity of the blastocyst.

9. Slowly expel the cells inside the blastocyst cavity (see Note 15).

10. Withdraw the pipette slowly.

11. Recover the blastocysts regularly and incubate them in drops of embryo culture medium under mineral oil in 35 mm dishes, within an incubator at 37 °C and 5 % CO_2 (see Note 16).

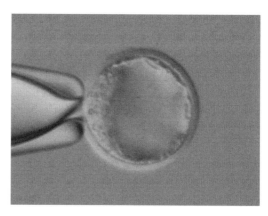

Fig. 2 Blastocyst positioned with the inner cell mass at 9 o'clock position, near the holding pipette (on the *left*)

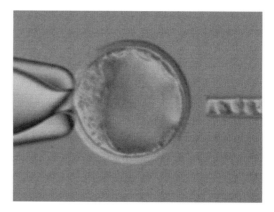

Fig. 3 Needle containing the cells for injection (on the *right*) approaching the surface of the blastocyst (on the *left*)

12. The injected blastocysts can be examined to see how many have collapsed and how many are beginning to reform blastocoelic cavities after the injection.

3.4 Uterine Blastocyst Transfer

1. Anesthetize 2.5 dpc pseudopregnant mice and place them on a pad.

2. Place a few small drops of buffered M2 medium into a 35 mm culture dish covered with mineral oil.

3. Take up a small amount of buffered medium into the embryo handling pipette, then load the blastocysts (we usually transfer seven to ten in each uterine horn) from the culture dish and expel into the loading dish droplets.

4. Take up a small amount of buffered medium into the handling pipette, then a small air bubble, and load up the blastocysts in a minimal volume of medium.

5. Place the embryo handling pipette aside (*see* **Note 17**).

6. Spray ethanol on the mice along the midline of the back.

7. Make a small incision (<1 cm) vertically along the midline of the back.

8. Slide the skin laterally to expose the fascia overlaying the left ovary and oviduct, the ovary can be seen alongside the fat pad.

9. Using forceps pick up the body wall covering the oviduct and cut through with the small scissors.

10. Using forceps grasp the fat pad attached to the oviduct and pull out the oviduct and upper part of the uterus.

11. Place the mouse on the dissecting microscope platform and maneuver the mouse in order to view the uterus through the eyepieces.

12. Pick up the embryo handling pipette and a 27G needle in the same hand.

13. Look for a site in the uterine wall, a few millimeters down from the uterotubal junction, clear of blood vessels. Use blunt forceps to hold the top of the uterus and make a hole with the 27G needle.

14. Withdraw the needle and insert the embryo handling pipette into the same hole.

15. Expel the embryos into the uterus until the air bubble moves down the pipette.

16. Withdraw the pipette slowly.

17. Reposition the skin incision over the right ovary/oviduct and repeat all the procedure.

18. Close the skin using sutures or wound clips.

19. Seventeen days after blastocyst transfer, full-term pups could be delivered either spontaneously or by abdominal surgery.

4 Notes

1. If the cells are growing on feeders it is necessary to remove the feeder cells. This can be done by plating the cells on gelatin for 30–60 min before starting the preparation of the cells.

2. According to the cells used different appropriate culture medium should be used.

3. During injection, the holding capillary needs to be controlled by an air micromanipulator control unit and the injection needle should be connected to a similar control unit filled with oil to provide better movement control.

4. To avoid cells to stick to each other and to the micromanipulation chamber bottom surface, the microscope stage should be cooled down to 20 °C.

5. The micromanipulation chamber can be made in-house, or can also be replaced with the cover of a 100 mm Petri dish.

6. Pipettes can be purchased or made in the laboratory. Holding pipettes are used to hold the blastocyst during the injection. Their quality contributes greatly to the success of blastocyst injection. The injection pipettes are used to pick up cells and inject them into the blastocyst.

7. For chimeras we recommend M2 and M16 or KSOM media. For 4n complementation we recommend H-CZB and CZB medium supplemented with 5.56 mM glucose.

8. If the cells were grown on feeder cells, the feeders will be loosely attached to the plastic surface, whereas the cells of interest will be in suspension. We therefore recommend placing the suspension in a tissue culture dish with gelatin and waiting for 1 h.

9. If the cells were grown on feeder cells, move the dish very carefully, taking care not to disturb the feeders attached to the dish, to collect the medium containing the cells of interest only.

10. The cells provided can be used for about 3 h.

11. For chimeras, it has been reported that embryos collected from superovulated females are not as good as those from naturally mated mice.

12. These mice can be obtained after a first intraperitoneal injection of 10 U PMSG and a second one with 10 U hCG 48 h later. The mice are then immediately mated with fertile males and the plugged females are selected on the following day (day 1).

13. We recommend M2 medium for chimeras and H-CZB for 4n complementation.

14. It is necessary to use well-expanded blastocysts, but not small or over-expanded ones as microinjection would then be more difficult.

15. Take care not to insert any oil or lysed cells into the blastocyst.

16. We recommend M16 or KSOM medium for chimeras and CZB medium (supplemented with 5.56 mM glucose) for 4n complementation.

17. Put the pipette well out of the way in order to avoid it from being knocked or dropped.

References

1. Tarkowski AK (1961) Mouse chimaeras developed from fused eggs. Nature 190: 857–860

2. Gardner RL (1968) Mouse chimeras obtained by the injection of cells into the blastocyst. Nature 220:596–597

3. Buehr M, Meek S, Blair K et al (2008) Capture of authentic embryonic stem cells from rat blastocysts. Cell 135:1287–1298

4. Zhao XY, Li W, Lv Z et al (2009) iPS cells produce viable mice through tetraploid complementation. Nature 461:86–90

Chapter 19

Whole-Mount In Situ Hybridization to Assess Advancement of Development and Embryo Morphology

Bernd Püschel and Alice Jouneau

Abstract

Whole-mount in situ hybridization (WISH) is widely used to visualize the site and dynamics of gene expression during embryonic development. Various methods of probe labeling and hybridization detection are available nowadays. Meanwhile the technique was adapted to be used on many different species and has evolved from a manual to a larger scale and automated procedure. Standardized automated protocols improve the chance to compare different experimental settings reliably. The high resolution of this method is ideally suited for examination of manipulated (e.g., cloned) embryos often displaying subtle changes only. Embedding and sectioning of in situ hybridized specimen further enhance the detailed examination of their gene expression and morphology.

Key words In situ hybridization, WISH, Methacrylate embedding, Sectioning

1 Introduction

Gene expression patterns within tissues or embryos can be detected using whole-mount in situ hybridization (WISH) resolving both the position and intensity of mRNA expression with high sensitivity and resolution. Temporal expression dynamics are accessible by analysis and comparison of consecutive developmental stages. Since WISH was first used in Drosophila embryos [1] it was successfully utilized to detect temporally and spatially restricted gene expression patterns in the embryonic development of many other species including domestic mammals (e.g., cow [2], rabbit [3, 4], sheep [5], pig [6], mouse [7]; see examples in Figs. 1 and 2). This knowledge is available as a reference assessing the development of cloned embryos and might help to understand the complex processes necessary to reprogram and redifferentiate a formerly somatic nucleus [8–10].

In vitro transcribed RNA probes are preferred for nonradioactive detection of cellular mRNAs because of their high labeling efficiency and temperature stability of the RNA–RNA hybrids

Nathalie Beaujean et al. (eds.), *Nuclear Reprogramming: Methods and Protocols*, Methods in Molecular Biology, vol. 1222, DOI 10.1007/978-1-4939-1594-1_19, © Springer Science+Business Media New York 2015

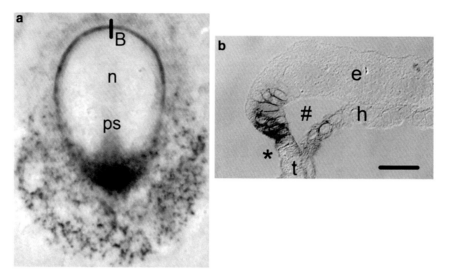

Fig. 1 BMP4 WISH of a rabbit embryo stage 4. En face view (**a**) of an embryonic disk oriented with the anterior pole to the top and a detail of a 5 μm Technovit section showing the anterior pole of the embryo at high magnification (**b**). BMP4 is expressed in an annular domain at the boundary of the embryonic disk, in the posterior third of the primitive streak and in the extraembryonic mesoderm (for further information see [22]. The section allows morphological identification of embryonic layers and structures and uncovers a clear BMP4 expression in the epiblast and a weak expression in the hypoblast at the boundary of the embryo. Asterisk marks the border of the embryonic disk, # label an artificial cleft between epiblast and hypoblast. Cell layers and structures are marked as follows: *e* epiblast, *h* hypoblast, *t* trophoblast, *n* Hensen's node, *ps* primitive streak. Bar 200 μm (**a**), 30 μm (**b**)

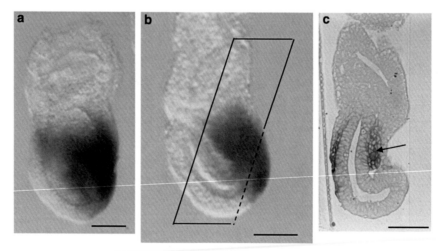

Fig. 2 LIM1 WISH of a mouse embryo at E7 (Mid-streak stage). LIM1 is expressed in the primitive streak and lateral mesodermal wings. (**a**) normal fertilized embryo, (**b**) cloned embryo with an abnormal, twisted shape and (**c**) Technovit section according to the plane indicated in **b**. The section uncovers the twist in the epiblast and confirms that mesoderm cells stained only, as expected. It also reveals an abnormal elongation of the streak, evidenced by the accumulation of cells in one side of the primitive streak (*arrow*). For further information, see [8]

resulting in a better signal to noise ratio. Depending on the labeling (digoxigenin or biotin) specifically hybridized riboprobes are detected by an appropriate antibody-coupled enzyme (alkaline phosphatase or horseradish peroxidase) converting a chromogenic substrate (e.g., BM Purple) into an insoluble color. Other variants of WISH use either fluorescently labeled probes or fluorescent dyes during detection to receive a higher resolution compared to enzymatic developed color (e.g., [11–13]). Here we describe a manual WISH protocol using a digoxigenin-labeled riboprobe as it was used in our recent work [14], which is customized for the use on different species and which can be adapted to an automated machine. For highest resolution of the three-dimensional distribution of the developed color and the overall embryonic morphology a protocol for methacrylate embedding and sectioning of embedded embryos is provided.

2   Materials

Use autoclaved ultrapure water (Millipore quality) and RNase- and DNase-free reagents to prepare all solutions. Set up a stock of Diethylpyrocarbonate (DEPC)-treated and autoclaved H_2O for the preparation of correspondingly labeled solutions. Take care that all containers need to be sterile and ribonuclease-free. Further methods, useful hints and the composition of solutions not described in detail here can be found in other standard molecular biology techniques collections [15].

2.1 Agarose gel Electrophoresis of the In Vitro Transcribed Probe

RNA-formamide loading buffer: 50 % (v/v) deionized formamide, 1× MOPS electrophoresis buffer (20 mM 3-[N-morpholino]propanesulfonic acid, pH 7.0, 5 mM sodium acetate, 1 mM EDTA).

Orange G loading dye: 20 % Ficoll, 10 mM Tris–HCl pH 7.5, 1 mg/ml Orange G.

2.2 Embryo Fixation

Fixation solution: 4 % PFA in DEPC-PBS, store in aliquots at –20 °C.

2.3 In Situ Hybridization Stock Solutions

Stock solutions are stored in aliquots at –20 °C.

1. Proteinase K, 1,000×: 10 mg/ml in DEPC-H_2O.

2. CHAPS, 20×: 10 % (w/v) in DEPC-H_2O.

3. Heparin, 1,000×: 50 mg/ml in DEPC-H_2O.

4. Yeast tRNA, 200×: 10 mg/ml in DEPC-H_2O.

5. Blocking reagent (BR), 10×: 10 % (w/v) Blocking reagent (Roche) in MAB.

6. Levamisol, 500×: 1 M in H_2O.

7. Normal goat serum (NGS): inactivate complement at 60 °C for 1 h.

2.4 In Situ Hybridization Buffers

1. DEPC-PBS: dilute 10× PBS with DEPC-treated H_2O.

2. PBT: 0.1 % Tween 20 in DEPC-PBS.

3. Methanol series: 75, 50, and 25 % (v/v) in PBT.

4. Hybridization buffer: 50 % Formamide, 1.4× SSC (*see* **Note 1**), 0.5 mM EDTA, 50 μg/ml tRNA (*see* **Note 2**), 0.2 % Tween 20, 0.5 % CHAPS, 50 μg/ml Heparin in DEPC-H_2O.

5. MAB: 100 mM Maleic acid, 150 mM NaCl, pH 7.5.

6. MABT: MAB containing 0.1 % Tween 20.

7. NTMT: 100 mM Tris, 100 mM NaCl, 50 mM $MgCl_2$, 2 mM Levamisol, 0.1 % Tween 20 (Prepare fresh just before use, because Levamisol is not stable.)

2.5 Metacrylate Embedding

1. Mowiol mounting medium.

2. Infiltration solution: 100 ml Technovit 8100 resin (basic solution) + 0.6 g hardener I; store in a dark bottle for up to 4 weeks at 4 °C.

3. Embedding medium: always freshly prepared: 6 ml infiltration solution + 250 μl hardener II.

2.6 Solutions for Section Counterstaining

1. Nuclear red or Eosin.

2. Toluidine Blue/Pyronin G: Mix 0.5 g of $Na_2B_4O_7 \cdot 10H_2O$ and 0.5 g Toluidine Blue in 50 ml H_2O. Mix 1 g Pyronin G in 100 ml. Then mix 3 volumes of Toluidine and 1 volume of Pyronin and use it as a 10× stock solution. Keep at room temperature in the dark.

2.7 Equipment

1. Blister pack, Blister foil.

2. Dry thermostat block (with holes snugly fitting the screw-cap polystyrene vials see below).

3. Mesh-bottomed baskets (Screw-cap cryo vials sealed with a small-meshed nylon netting; [16]).

4. Platform rocker.

5. Screw-cap polystyrene vials (7 ml Bijou container).

6. Tissue culture plates (24 well).

7. Thermocycler.

   Further reagents and equipment are mentioned in the text where appropriate.

3 Methods

3.1 RNA Probe Synthesis

Clone the cDNA of interest to a plasmid containing a multiple cloning site flanked by SP6-, T3- or T7-promoter sequences (e.g., pGEM T-easy or pBluescript II). A suitable template for the in vitro transcription reaction can be generated either by linearization of the plasmid DNA using an appropriate restriction enzyme [15] or by PCR amplification of a plasmid fragment [17]. The choice of the restriction enzyme, respectively the primer combination will generate the intended orientation (sense or antisense) of the cloned cDNA which is flanked at its 5′-end by one of the aforementioned promoter sequences.

1. Purify the template with a PCR Purification column.

2. Use an in vitro transcription kit to perform the RNA probe synthesis as follows:

cDNA template (1 μg linearized plasmid or 200 ng PCR product)	NB: x μl
10× transcription buffer	2 μl
10× DIG-labeling mix	2 μl
RNase free H$_2$O up to 18 μl	NB: x μl
SP6/T7/T3-RNA polymerase (20 U/μl)	2 μl

3. Incubate at 37 °C for 2 h.

4. Stop the reaction by addition of 2 μl 0.2 M EDTA (pH 8.0).

5. Remove unincorporated nucleotides either by two consecutive ethanol precipitations (using 0.1 volumes of 4 M LiCl and 2.5 volumes of prechilled (−20 °C) ethanol; dissolve the final probe in 25 μl RNase-free water) or by purification on a Quick Spin Column for RNA purification.

6. Add 1 μl RNase inhibitor.

7. To verify success of the synthesis mix a 1 μl aliquot of the probe with 9 μl RNA-formamide loading buffer, incubate at 65 °C for 5 min, chill on ice, add 3 μl Orange G loading dye and analyze the sample on a 1–2 % TBE-agarose gel.

8. Store the probe at −20 °C or at −80 °C for long-term storage.

3.2 Embryo Preparation

Early non-attached embryos are collected either by gently flushing the fallopian tubes or the uterus with warm PBS. Attached embryonic stages are extracted from each implantation chamber separately (pig [6], rabbit [18], bovine [2, 10, 19], mouse [20, 21]). Isolated embryos are rinsed in fresh PBS and further treated as described below:

1. Fix the embryos in 4 % paraformaldehyde (DEPC-PBS) for 1–2 h (according to the size and embryonic stage) at room temperature.

2. Rinse 3 times with DEPC-PBS for 5 min each at room temperature.

3. Dehydrate the embryos for 5 min at each step with an ascending series of ice-cold methanol solutions in PBS: 25, 50, 75, 100 %.

4. After one additional wash in fresh methanol store the embryos in 100 % methanol at –20 °C until needed.

3.3 Hybridization and Riboprobe Detection

Embryos can be processed either manually in self-made mesh-bottomed baskets [16] or automatically with a robot (InsituPro VS, Intavis; *see* **Note 3**). Manual execution of the incubations can be conveniently performed, e.g., in 24-well culture plates (room temperature, 4 °C) and screw-cap polystyrene vials (70 °C) reducing the risk of drying of the embryonic tissue during the transfer of the mesh-bottomed baskets from one solution to the next. Depending on the number of probes and the number of washes, one can calculate the necessary total buffer volume based on a volume of 1.2 ml per well. Volumes and amounts mentioned in the description reflect typical values for a single riboprobe processed with our self-made mesh-bottom baskets. Include a sense probe hybridization to control for nonspecific signal and overall level of background staining. If possible also append a positive control, e.g., a riboprobe known to hybridize well at the embryonic stage of interest to check for the quality and reliability of all steps of the procedure.

1. Rehydrate the embryos on ice for 5 min each step using a decreasing methanol-PBT-series containing 75, 50, and 25 % (v/v) methanol.

2. Wash 3 times for 5 min with PBT on ice.

3. Incubate the embryos in 10 μg/ml proteinase K in PBT at room temperature. The digestion time depends on the species, the stage of the embryo, and sometimes need to be optimized according to the gene expression to be detected. See Table 1 for examples of incubation.

4. Wash 2 times for 5 min in PBT at room temperature.

Table 1
Examples of proteinase K incubation times for different mammalian embryos

Time (min)	0	5	10	20
Rabbit		Stage 0–3	Stage 4–6	Stage 6–8
Bovine	Up to day 7		Day 8–17	
Mouse	Up to E6.5	E7-8.5	E9-11	
Pig	Up to E6		Day 7–12	

NB: except for mouse embryos, all others are dissected prior to the WISH in order to leave only the embryonic disk and a piece of extra-embryonic tissue around. Do not use the whole blastocyst, as its cavity would capture some unspecific staining

5. Fix the embryo in 0.2 % glutaraldehyde in PBT for 20 min at room temperature.

6. Set the aluminum dry thermostat block to 70 °C and warm for each probe two separate screw-cap polystyrene tubes each filled with 900 µl hybridization buffer at 70 °C.

7. Optional: Incubate 10 min in Glycine solution (4 mg/ml in PBT) after the post-fixation (**step 5**) to block remaining free aldehyde radicals which may otherwise cause unspecific binding of antibodies.

8. Wash the fixed embryos twice in PBT, once in a 50:50 mixture of PBT and hybridization buffer and once in 100 % hybridization buffer (for 5 min at room temperature).

9. Transfer the embryos into the preheated hybridization buffer (first screw-cap polystyrene tube) and allow prehybridization to occur for at least 1 h.

10. Add the digoxigenin-labeled riboprobe (1 µg) to 100 µl hybridization buffer and denature at 95 °C for 5 min (Thermocycler).

11. Add the denatured probe into the second preheated aliquot (second screw-cap polystyrene tube, 70 °C), mix with a short swirl and immediately transfer the embryo into this hybridization buffer containing the riboprobe. Hybridize overnight (16 h) at 70 °C (*see* **Note 4**).

12. The next day prepare five separate aliquots (1 ml each) containing hybridization buffer and one aliquot (1 ml) containing a 50:50 mixture of hybridization buffer and MAB and warm them to 70 °C (*see* **Note 5**).

13. Unbound riboprobe is removed by consecutive incubation at 70 °C in hybridization buffer for 1 min (2×) and 30 min (3×) and in the hybridization buffer/MAB mixture for 15 min.

14. Alternative post-hybridization washings: 30 min (3×) in formamide 50 %/SSC 2× at 70 °C, 30 min (3×) in SSC 2×/Tween 0.1 % at 55 °C, and then 30 min (3×) in SSC 0.2×/Tween 0.1 % at 55 °C. This alternative may reduce background, if any.

15. Wash 3 times in MABT and incubate 2 times in MABT for 30 min each, both at room temperature.

16. Block the samples at room temperature in MABT containing 2 % BR for 1 h and in MABT containing 2 % BR and 20 % NGS for an additional hour.

17. Replace the last blocking solution by a fresh aliquot of the same composition, additionally containing a 1/2,000 diluted alkaline phosphatase coupled anti-DIG antibody. Incubate overnight at 4 °C.

18. Unbound antibody is removed by successive washes with MABT for 1 min (3×) and 30 min (7×) at room temperature.

19. Incubate the samples 4 times in NTMT, 10 min each at room temperature.

20. Remove the embryos from the mesh-bottomed baskets and transfer them to a petri dish containing an appropriate precipitating alkaline phosphatase substrate (e.g., BM purple, Roche).

21. Incubate in a humidified chamber in the dark at room temperature for several minutes up to some days. The appearance and the speed of the color development depend on several factors such as the level of the gene expression, the pretreatment of the embryo, or the choice of the probe. Therefore regularly check the staining under the stereomicroscope, as it can emerge quickly. To pause the reaction (e.g., overnight) wash and store the embryo in NTMT and resume the reaction, e.g., the next day by returning the embryo back in the alkaline phosphatase substrate solution (*see* **Note 6**). During the reaction do not expose the samples to light for prolonged periods of time.

22. To stop the reaction, wash the embryo several times with NTMT and transfer the sample to PBS.

23. Take photographs and store the embryos in glycerol or embed them in Technovit (*see* Subheading 3.4) for subsequent detailed examination of the gene expression on thin sections.

24. Results: Whole-mount in situ hybridizations of a rabbit and a mouse embryo are shown in Figs. 1a and 2a, b.

3.4 Embedding in Technovit 8100

1. Optional: the hybridized embryo might be flattened overnight by gently spreading in Mowiol mounting medium between a glass slide and a coverslip. The next day incubate the glass slide for 1 h in 90 % Methanol. Gently detach the coverslip and recover the embryo in 100 % Methanol. Continue with **step 3**.

2. Dehydrate non-spread embryos using an ascending methanol-PBS series (PBS containing 25 %, 50 %, 75 % volume methanol) for 15 min each at room temperature.

3. Transfer the embryos in fresh 100 % Methanol and leave them for 15 min at room temperature.

4. Precool a metal block in a wet ice bath (to gain optimal temperature exchange during the following steps) and prepare the infiltration solution.

5. Move the embryos to the infiltrations solution (4-well culture plate) and incubate on the metal block for 90–120 min.

6. Set up the embedding medium in chilled plastic molds (e.g., blister foil) on the metal block. Solidified embedding medium from earlier embedding procedures may be cut to size and may

serve as a support preventing the embryo from sedimentation to the bottom of the plastic mold.

7. Transfer the embryos to the embedding medium, adjust the desired position and seal the opening with plastic foil without trapping any air inside the plastic mold.

8. Carefully move the ice box with the metal block and the assembled plastic mold in a refrigerator (4 °C) and allow polymerization overnight. Prolonged storage of polymerized Technovit blocks at room temperature should be avoided (blocks tend to get brittle, making the following preparation of sections difficult; in the worst case the integrity of the tissue will be affected). Storage up to 4 weeks at 4 °C or up to 1 year at –20 °C prevents or at least slows down these changes of the Technovit and makes the subsequent processing more convenient.

3.5 Sectioning

1. For sectioning take out the Technovit block from the plastic mold and incubate in a water bath at 70 °C until the plastic is getting more flexible.

2. Use a razor blade to remove excess embedding material until the desired shape is acquired.

3. Generate serial sections of 5 μm thickness on a microtome using glass knifes.

4. Transfer each section to a separate 50 μl of a 50:50 water-ethanol mixture on a glass slide.

5. Subsequent drying at 37 °C will spread the section optimally on the glass slide.

6. Examine the sections using a microscope preferably with Nomarski contrast to visualize the unstained embryonic tissue as convenient as the in situ hybridization signals.

3.6 Counterstaining

If counterstaining is desired, different stains can be used:

1. Nuclear red, Eosin: cover the sections with the solution and leave for 5–15 min, depending on the desired intensity of staining. Then remove the stain and rinse quickly in distilled water. If the ISH staining is relatively weak, be careful not to mask it with the pink/red counterstaining.

2. Toluidine Blue/Pyronin G: Dilute the 10× stock solution in water and put one drop of it on the sections, leave 30 s before rinsing in distilled water. This gives a strong blue staining, so do not use it on all sections, as the ISH staining will be masked.

3. After staining dry the slides and immediately mount the sections using a permanent, non-aqueous medium such as Eukitt.

4. Results: such sections are shown in Figs. 1b and 2c.

4   Notes

1. pH of SSC: acid pH (5.5) slightly increases the stringency of the hybridization procedure but may also partially hydrolyze the probe. So pH 7 should be preferred for the hybrid mix, whereas it is possible to use pH 5.5 for the subsequent washings, if background is high.

2. tRNA in the hybridization mixture: can be omitted, but may reduce background, if any.

3. If using InSituProVS robot, utilize medium baskets; the size of the mesh will depend on the size of the embryos. The exchange volume is 700 µl. It can be lowered to 400 µl for mouse embryos up to E7.5, bovine up to day 14, pig up to day 10, and rabbit stage 3.

4. Hybridization temperature can be set at 65 °C instead of 70 °C in case the probe does not exactly match the target sequence (for example for cross-species hybridization): up to 85 % identity between the probe sequence and the target RNA sequence might work well.

5. Keep the hybridization buffer containing the probe (−20 °C or −80 °C). It can be reused up to two times.

6. If whole-mount in situ hybridized embryos shall be used for histological sectioning, develop the specific staining as intensely as possible, while ensuring that the background does not come up as well.

Acknowledgments

The authors' research has been supported in part by grants as specified in our publications and by INRA and the University of Göttingen. We gratefully appreciate our colleagues for inspiring and helpful input (in Göttingen: Christoph Viebahn; in Jouy: Evelyne Campion) and for their excellent and dedicated help generating nice WISH data (in Göttingen: Kirsten Falk-Stietenroth, Heike Faust; and Irmgard Weiss, in Jouy: Vincent Brochard).

References

1. Tautz D, Pfeifle C (1989) A non-radioactive in situ hybridization method for the localization of specific RNAs in Drosophila embryos reveals translational control of the segmentation gene hunchback. Chromosoma 98(2):81–85

2. Hue I, Renard JP, Viebahn C (2001) Brachyury is expressed in gastrulating bovine embryos well ahead of implantation. Dev Genes Evol 211(3):157–159

3. Idkowiak J, Weisheit G, Plitzner J, Viebahn C (2004) Hypoblast controls mesoderm generation and axial patterning in the gastrulating rabbit embryo. Dev Genes Evol 214(12):591–605

4. Idkowiak J, Weisheit G, Viebahn C (2004) Polarity in the rabbit embryo. Semin Cell Dev Biol 15(5):607–617

5. Guillomot M, Turbe A, Hue I, Renard JP (2004) Staging of ovine embryos and expression

of the T-box genes Brachyury and Eomesodermin around gastrulation. Reproduction 127(4):491–501

6. Hassoun R, Schwartz P, Feistel K, Blum M, Viebahn C (2009) Axial differentiation and early gastrulation stages of the pig embryo. Differentiation 78(5):301–311

7. Herrmann BG (1991) Expression pattern of the Brachyury gene in whole-mount TWis/TWis mutant embryos. Development 113(3): 913–917

8. Jouneau A et al (2006) Developmental abnormalities of NT mouse embryos appear early after implantation. Development 133(8):1597–1607

9. Rielland M, Brochard V, Lacroix MC, Renard JP, Jouneau A (2009) Early alteration of the self-renewal/differentiation threshold in trophoblast stem cells derived from mouse embryos after nuclear transfer. (Translated from eng). Dev Biol 334(2):325–334

10. Degrelle SA et al (2012) Uncoupled embryonic and extra-embryonic tissues compromise blastocyst development after somatic cell nuclear transfer. PLoS One 7(6):e38309

11. Nakamura T, Hamada H (2013) Fluorescent 2 color Whole mount in situ hybridization for a mouse embryo. Protocol Exchange doi:10.1038/protex.2013.002

12. Acloque H, Wilkinson DG, Nieto MA (2008) In situ hybridization analysis of chick embryos in whole-mount and tissue sections. Methods Cell Biol 87:169–185

13. Brend T, Holley SA (2009) Zebrafish whole mount high-resolution double fluorescent in situ hybridization. J Vis Exp doi: 10.3791/1229

14. Püschel B, Blum M, Viebahn C (2010) Whole-mount in situ hybridization of early rabbit embryos. Cold Spring Harb Protoc (1):pdb prot5355

15. Sambrook J, Russell DW (2001) Molecular cloning: a laboratory manual, 3rd edn. Cold Spring Harbor Laboratory Press, Cold Spring Harbor, NY

16. Weisheit G, Mertz D, Schilling K, Viebahn C (2002) An efficient in situ hybridization protocol for multiple tissue sections and probes on miniaturized slides. Dev Genes Evol 212(8): 403–406

17. Hue I, Degrelle SA, Viebahn C (2013) Analysis of molecular markers for staging perigastrulating bovine embryos. Methods Mol Biol 1074:125–135

18. Püschel B, Viebahn C (2010) Rabbit mating and embryo isolation. Cold Spring Harb Protoc (1):pdb prot5350

19. Degrelle SA et al (2005) Molecular evidence for a critical period in mural trophoblast development in bovine blastocysts. Dev Biol 288(2):448–460

20. Hogan B, Beddington R, Costantini F, Lacy E (1994) Manipulating the mouse embryo: a laboratory manual, 2nd edn. Cold Spring Harbor Laboratory Press, Cold Spring Harbor, NY

21. Shimizu H, Uchibe K, Asahara H (2009) Large-scale whole mount in situ hybridization of mouse embryos. Methods Mol Biol 577: 167–179

22. Hopf C, Viebahn C, Püschel B (2011) BMP signals and the transcriptional repressor BLIMP1 during germline segregation in the mammalian embryo. Dev Genes Evol 221(4):209–223

Chapter 20

Genome-Wide Analysis of Methylation in Bovine Clones by Methylated DNA Immunoprecipitation (MeDIP)

Hélène Kiefer

Abstract

Methylated DNA immunoprecipitation (MeDIP), when coupled to high-throughput sequencing or microarray hybridization, allows for the identification of methylated loci at a genome-wide scale. Genomic regions affected by incomplete reprogramming after nuclear transfer can potentially be delineated by comparing the MeDIP profiles of bovine clones and non-clones. This chapter presents a MeDIP protocol largely inspired from Mohn and colleagues (Mohn et al., Methods Mol Biol 507:55–64, 2009), with PCR primers specific for cattle, and when possible, overviews of experimental designs adapted to the comparison between clones and non-clones.

Key words Cattle, Clones, Nuclear transfer, DNA methylation, MeDIP, Microarray, Epigenomics, CpG, Genome-wide analysis

1 Introduction

DNA methylation defaults have been reported in bovine clones, both at the global level [1, 2] and at candidate loci [3, 4], but a genome-wide view of the affected regions is still lacking. Several methods are available to perform such an analysis. Single-base resolution is achieved using techniques based on the conversion of unmodified cytosine to uracil by sodium bisulfite, followed by next-generation sequencing. The cost of whole genome bisulfite-sequencing (WGBS; [5, 6]) still precludes the analysis of a large number of animals. Reduced representation bisulfite-sequencing (RRBS; [7]) is much more affordable, but targets CpGs in a particular sequence context (moderate to high CpG density). Furthermore, WGBS and RRBS cannot discriminate DNA methylation from hydroxymethylation [8, 9]. Since these two marks may have different functions both in the somatic nucleus and during reprogramming to totipotency, it could be beneficial to distinguish between them when analyzing the epigenome of clones. Recently, bisulfite-based approaches have been developed to map

Nathalie Beaujean et al. (eds.), *Nuclear Reprogramming: Methods and Protocols*, Methods in Molecular Biology, vol. 1222, DOI 10.1007/978-1-4939-1594-1_20, © Springer Science+Business Media New York 2015

hydroxymethylation at a genome-wide scale and single-base resolution [10, 11]. However, these techniques require comparison with traditional bisulfite-sequencing profiles, which precludes their large scale utilization for the mapping of epigenetic alterations in bovine clones.

Techniques using the capture of methylated DNA by methyl-binding proteins [12–14] or by anti-methylcytidine antibody (MeDIP; [15, 16]) are good alternatives to bisulfite-based methods. Although showing less resolution than bisulfite-based methods [17, 18], capture-based methods are specific for methylated DNA [8] and allow a variety of subsequent applications. Enrichment of candidate regions in the captured fraction can be estimated using PCR or quantitative PCR, and captured sequences can be identified at a genome-wide scale by sequencing or by hybridization on a microarray. These capture-based techniques are still widely used to analyze the human methylome [19].

The method described here has been developed for MeDIP followed by microarray hybridization (MeDIP-chip), but can also be adapted to MeDIP-sequencing (MeDIP-seq), with some modifications (*see* **Note 8** and protocols in [20]). In contrast to MeDIP-chip, MeDIP-seq allows to explore the whole methylome and to identify regions of interest outside of CpG islands and promoters. We analyze different tissues from bovine clones and non-clones using a cost-effective strategy which combines both MeDIP-seq and MeDIP-chip. We take advantage of MeDIP-seq profiles generated on a restricted number of animals to design a custom microarray, which in turn can be used to interrogate the methylome of a larger cohort. Results are then validated using bisulfite-sequencing (Chapter 18) or pyrosequencing on candidate regions.

2   Materials

2.1   Purification of Genomic DNA

1. Lysis buffer: 50 mM NaCl, 10 mM Tris–HCl pH 7.5, 10 mM EDTA, 0.2 % SDS.

2. Proteinase K: resuspend in H_2O and store aliquots at −20 °C.

3. RNAse A: resuspend in H_2O and store aliquots at −20 °C.

4. Tris-saturated phenol pH 8.0.

5. TE (Tris–EDTA) buffer: 10 mM Tris–HCl pH 8.0; 1 mM EDTA pH 8.0.

6. Water bath, incubators, bowl and pestle.

2.2   Sonication of Genomic DNA

1. S220 Focused-Ultrasonicator (Covaris), including the temperature-monitoring device and the circulation pump.

2. Tube and Cap TC13 13×65 mm (Covaris #520010) and holder.

3. 50 % Glycerol in H_2O.

2.3 MeDIP

1. 1 M Na-phosphate buffer pH 7.0: 42.3 mL 2 M monobasic sodium phosphate (NaH$_2$PO$_4$), 57.7 mL 2 M dibasic sodium phosphate (Na$_2$HPO$_4$), 100 mL H$_2$O.

2. 10× IP buffer: 1.4 M NaCl, 100 mM Na-phosphate buffer pH 7.0, 0.5 % Triton X-100.

3. 1× IP buffer: one volume 10× IP buffer and nine volumes H$_2$O.

4. Anti 5-methylcytidine antibody: we use Eurogentec 5-methylcytidine antibody (BI-MECY-1000, 1 mg/mL), but different sources have been successfully used by others. It may be necessary to adjust the amount of antibody when using a different source.

5. Magnetic beads: we use Dynabeads M-280 sheep anti-mouse IgG.

6. BSA/PBS buffer: 1 mg/mL BSA (molecular biology grade) resuspended in 1× PBS.

7. Proteinase K: resuspend at 10 mg/mL in H$_2$O and store aliquots at –20 °C.

8. PK digestion buffer: 50 mM Tris–HCl pH 8, 10 mM EDTA, 0.5 % SDS.

9. Magnetic rack and tube rotator.

2.4 Quality Control of MeDIP Samples

1. Cattle-specific primer sets (Table 1).

2. Taq polymerase: the amplification conditions we describe here were obtained using Platinum DNA Taq polymerase (Invitrogen #10966083). Any other Taq polymerase can probably be used, but the conditions may have to be adjusted.

3. dNTP mix.

4. Thermal cycler.

2.5 Whole Genome Amplification (WGA)

1. Amplification: GenomePlex complete WGA kit (Sigma-Aldrich #WGA2-50RXN).

2. Purification: we use the QIAquick PCR purification kit (Qiagen #28104).

3 Methods

3.1 General Considerations About the Experimental Design

Previous work has demonstrated that adult bovine clones show more variability than monozygotic twins in the global content of methylcytosine, independently of their genotype [1]. This variability should be taken into account in the experimental design, which should include enough animals especially in the "clone" category. To our experience, a sample size of five clones is enough to identify cloning-associated epigenetic signature, however with limited statistical power.

Table 1
Cattle-specific primers for quality control of the MeDIP samples by PCR

Name	Sequence (5′-3′)	Size of the product	Hybridization temperature	Number of cycles	Description
bDAZL_F2 bDAZL_R2	CAAGCACTTCACTTCTCCAACA CCCACTGATTGGACTGTTTGT	210 bp	58 °C	35	Sequence without CpG
bH19_F1 bH19_R1	CGAGGGGTACTGAGAGGTTG TCCTCTCCCACCTTCAACAG	242 bp	58 °C	40	Imprinted region (putative CTCF site)
bH19_F2 bH19_R2	GCAGGAGGATTTCACAG CACCTTAGGAGGCTCAGACG	217 bp	58 °C	40	Imprinted region (putative CTCF site)
bH19_F3 bH19_R3	ACAGCCCTAGCTCCAGAGG AATCAGACCTCAGGCACAGC	241 bp	58 °C	40	Imprinted region (putative CTCF site)
bIGF2_F1 bIGF2_R1	CAGAGAGGCCAAGAGTCACC AGGACGGTACAGGGATTTCA	203 bp	58 °C	35	IGF2 intragenic DMR [26, 27]
bPEG3_F1 bPEG3_R1	CCTAGCTGCGTTTCACAGGT AACAAACTGCAGTGGGAAGG	239 bp	58 °C	35–40	PEG3 upstream DMR [28]
bSNRP_F1 bSNRP_R1	GTAGACACCATCCCGGTTT AACACAAGTCACGCATCTC	221 bp	58 °C	35	SNRPN promoter DMR [29]
bSNRP_F2 bSNRP_R2	CAGCAAGAGTTGGTAAAGTATCTGA TGCCTGCCATTCAATCTGT	159 bp	58 °C	35	Sequence without CpG
bTBX15_F1 bTBX15_R1	AGTCTACACCGCCTGTGAGC GAGGCGGCCAAGACTGAG	198 bp	58 °C	35	Housekeeping gene promoter/exon 1, CpG island

These primers target differentially methylated regions (DMR) which should be detected in the MeDIP samples (H19, IGF2, PEG3, bSNRP_F1 × R1), or unmethylated CpG-rich promoters (TBX15) and regions devoid of CpG (bSNRP_F2 × R2, bDAZL_F2 × R2), none of which should be detected in the MeDIP samples. All the PCR products should be detected in the input samples

Another point which should be considered when studying the consequences of nuclear reprogramming on the epigenome is the nature of the control sample (non-cloned animals). Depending on the scientific issue, this control sample could consist in animals obtained by conventional artificial insemination or in vitro fertilization, or in monozygotic twins. In all these cases, it could be beneficial to introduce genetic diversity in the clone group (i.e. clones with different genotypes), to avoid confounding effects of genotypes and nuclear reprogramming. Another alternative is to compare genetically identical clones with their cell donor.

3.2 Purification of Genomic DNA

1. Tissue collection: using forceps and scalpels, quickly cut the tissue into small pieces (about 0.5 cm long; *see* **Note 1**). Snap freeze the pieces one by one into liquid nitrogen and store at –80 °C until use.

2. Prepare the appropriate amount of lysis buffer containing 0.2 mg/mL proteinase K. We use 10 mL proteinase K-containing buffer per gram tissue. One piece of tissue (0.2–0.3 g) is usually enough to get high amounts of good quality DNA.

3. Wearing insulating gloves, grind the tissue in liquid nitrogen or dry ice.

4. Transfer the frozen powder to the tubes containing lysis buffer and proteinase K. Mix immediately by inverting the tube several times (*see* **Note 2**).

5. Incubate overnight at 55 °C.

6. Add 25 μg/mL RNAse A and incubate for 1 h at 37 °C.

7. Add 0.2 mg/mL proteinase K and incubate for 1 h at 55 °C.

8. Extract with one volume phenol–chloroform (1:1). Centrifuge several minutes at low speed and transfer the upper phase to a new tube.

9. Extract with one volume chloroform. Centrifuge several minutes at low speed and transfer the upper phase to a new tube.

10. Precipitation: add 0.2 M NaCl to the sample and chill on ice. Slowly add one volume ethanol and mix by inverting the tube several times. Let sit a few minutes on ice (*see* **Note 3**).

11. Using a Pasteur pipette, transfer the precipitate of genomic DNA to a tube containing 75 % ethanol. Mix gently.

12. Air-dry genomic DNA a few minutes on the Pasteur pipette upside-down, and transfer to a tube containing TE. Gently resuspend (*see* **Note 4**).

13. Measure concentration and check the quality of genomic DNA on an agarose gel (*see* **Note 5**).

3.3 Sonication of Genomic DNA

The following protocol has been optimized to generate 100–1,200 bp-fragments using a S220 Focused-Ultrasonicator (Covaris). Protocols using other devices are described in [15].

1. Dilute 10–200 μg genomic DNA in 300 μL TE (*see* **Note 6**).

2. Add 200 μL of 50 % glycerol and transfer to a TC13 tube. Save 5–10 μL to run on an agarose gel (Fig. 1, lanes 1–3).

3. Using the Covaris software, program the following parameters: duty cycle 2 %; intensity 7; cycle/burst 200; 65 cycles (10 s/cycle).

4. Place the tube on the holder and sonicate while the tube is cooled in the water bath. The program will last approximately 10 min per sample.

5. Load 5–10 μL on a 1.5 % agarose gel. After sonication, DNA should migrate as a smear with an average size of 450 bp (Fig. 1, lanes 4–6).

6. Transfer sonicated DNA to a 2 mL microtube. Add 400 mM NaCl (40 μL 5 M NaCl in 500 μL DNA), 1 μL glycogen (5 μg/μL), and 1 mL ethanol. Incubate for at least several hours at –20 °C.

7. Centrifuge for 1 h at $18,000 \times g$ in a refrigerated centrifuge. Wash pellet with 75 % ethanol, dry pellet and resuspend in 50–100 μL TE. Measure DNA concentration and store at –20 °C.

3.4 MeDIP

For the organization of the experiment, *see* Fig. 2.

MeDIP leads to the enrichment of methylated DNA by immunoprecipitation using an anti 5-methylcytidine antibody. For applications like PCR and microarray hybridization, the degree of enrichment is estimated by comparison with the input DNA (sonicated DNA which is left untreated). Therefore, it is important to save enough input material for subsequent analysis.

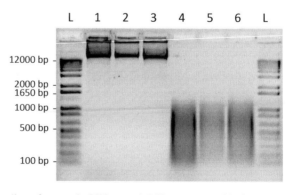

Fig 1 Migration of genomic DNA on a 1.5 % agarose gel before (*lanes 1–3*) and after (*lanes 4–6*) sonication. L, DNA ladder

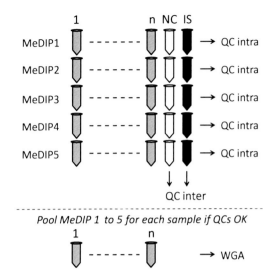

Fig 2 Organization of the MeDIP experiments. MeDIP1 to MeDIP5: technical repetitions; 1-n, samples (tubes in *grey*); *NC* MeDIP negative control without antibody (10 μL antibody replaced by 10 μL H$_2$O; *white*), *IS* internal standard (MeDIP performed with the same sample for all technical repetitions; *black*), *QC intra* intra-experiment quality control (duplex PCR with bIGF2_F1 × R1 and bSNRP_F2 × R2), *QC inter* inter-experiment quality control performed on NC and IS (gel migration, PCR with other primers)

The remaining sonicated DNA can be divided into 4 μg-aliquots, which will be used to perform several independent MeDIP repetitions per animal. These repetitions are necessary to avoid technical biases. We typically perform five technical repetitions, which we pool before proceeding to whole-genome amplification/microarray hybridization. If the amount of starting material is limiting, the best alternative is to use an automated device (for instance SX-8G IP-Star, Diagenode), which limits technical biases (personal communication from J. Tost, CNG, Evry, France). In these conditions, a unique MeDIP reaction can be considered as sufficient per sample. If such device is not available, miniaturization protocols can be found in [21].

In addition, using a sample for which the amount of material is not limiting, include a negative control without antibody (replace 10 μL antibody by 10 μL H$_2$O, all subsequent treatments being unchanged), and an antibody-containing internal standard. This internal standard, produced from the same animal for all technical repetitions, will not be hybridized on microarrays. It will be used to compare the efficiency of all technical repetitions before pooling (*see* Subheading 3.5, Quality control of MeDIP samples, and Fig. 2).

1. Dilute 4 μg sonicated DNA in 450 μL TE (*see* **Note 7**).

2. Denature for 10 min at 100 °C and immediately cool on ice at least 10 min. It is necessary to obtain and conserve single-strand

structures for optimal interaction with the anti 5-methylcytidine antibody. For MeDIP-seq applications, *see* **Note 8**.

3. On ice, add 51 μL 10× IP buffer. Mix by pipetting up and down several times.

4. On ice, add 10 μL anti 5-methylcytidine antibody (10 μg). Replace by 10 μL H₂O in the negative control tube. Incubate for 3–4 h at 4 °C with constant shaking using a tube rotator. Continue at **step 11**.

5. During incubation with the antibody, prepare the magnetic beads (*see* **Note 9**). Transfer 40 μL bead suspension to a microtube. If n samples are treated in parallel, transfer n× 40 μL to a microtube.

6. Using a magnetic rack, trap the beads on the wall of the tube. When the supernatant becomes clear, carefully aspirate it without disturbing the beads.

7. Resuspend the beads in 800 μL BSA/PBS buffer per sample. Incubate for 5 min at room temperature with constant shaking to prevent sedimentation.

8. Trap the beads using the magnetic rack and remove supernatant.

9. Repeat once the washing step with BSA/PBS.

10. Resuspend the beads in 40 μL of 1× IP buffer per sample.

11. Add the beads to the samples (*see* **Note 9**) and incubate for 2 h at 4 °C with constant shaking.

12. Trap the beads using the magnetic rack and remove supernatant (*see* **Note 10**).

13. Resuspend the beads in 800 μL of 1× IP buffer and incubate for 5–10 min at room temperature with constant shaking (*see* **Note 11**).

14. Trap the beads using the magnetic rack and remove supernatant.

15. Repeat twice the washing step with 800 μL of 1× IP buffer.

16. When the third washing is completed, remove 1× IP buffer and resuspend the beads in 250 μL PK digestion buffer (*see* **Note 12**).

17. Add 7 μL of 10 mg/mL proteinase K and incubate for 3 h at 50 °C with constant shaking (*see* **Note 13**).

18. Extract with 250 μL phenol and collect the aqueous phase.

19. Extract with 250 μL chloroform and collect the aqueous phase.

20. Precipitation: add 400 mM NaCl (20 μL of 5 M NaCl in 250 μL DNA), 1 μL glycogen (5 μg/μL), and 500 μL ethanol. Incubate for at least several hours at −20 °C.

21. Centrifuge for 1 h at 18,000 × *g* in a refrigerated centrifuge (*see* **Note 14**). Wash pellet with 75 % ethanol, dry pellet, and resuspend in 6.2 µL H₂O (0.2 µL will be used for systematic PCR check; *see* Subheading 3.5, **step 2**). Store at −20 °C.

3.5 Quality Control of the MeDIP Samples

After each MeDIP experiment, the presence of methylated DNA and the absence of unmethylated DNA in the samples can be easily checked using the duplex PCR described in steps 1–8 (Fig. 2, "QC intra"). Additional controls can be performed using the internal standards and negative controls included in each technical repetition (**steps 9** and **10** and Fig. 2, "QC inter").

1. Prepare a mix containing: 1× PCR reaction buffer; dNTP mixture (200 µM each); 1.5 mM MgCl₂; the four following primers: bIGF2_F1, bIGF2_R1, bSNRP_F2, bSNRP_R2 (0.2 µM each; see Table 1), 1 unit Platinum Taq polymerase in 50 µL final volume per sample. Prepare enough mix to include one control without template and one input sample (mix for n samples + negative control + internal standard + 2 tubes; *see* **Note 15**).

2. Assemble the PCR reaction in 0.2 mL PCR tubes or strips: 50 µL mix + 0.2 µL MeDIP sample or H₂O or 20 ng input.

3. Cap the tubes, vortex briefly, and spin down to collect content.

4. Using a thermal cycler, denature for 3 min at 94 °C.

5. Perform 35 cycles of PCR amplification as follows: 30 s at 94 °C, 1 min at 58 °C, 1 min at 72 °C.

6. Incubate for 10 min at 72 °C.

7. Run 5 µL of the products on a 2 % agarose gel (*see* **Note 16**).

8. Interpretation (Fig. 3): the two PCR products should be detected in the input sample, but only the IGF2 PCR product should be detected in the MeDIP samples (upper band). No band should be observed in the MeDIP negative control without antibody or in the PCR negative control without template.

9. Additional PCR can be performed using the internal standards as templates, to check the reproducibility of the MeDIP technical repetitions. Cattle-specific primers and PCR conditions are listed Table 1.

10. Migrate 5 µL of internal standards and negative controls without antibody on a 1.5 % agarose gel. Internal standards should be devoid of degradation (migration pattern similar to Fig. 1, lanes 4–6), and should show homogenous amounts of material between MeDIP technical repetitions. No DNA should be observed in the negative controls.

11. Pool the technical repetitions that have successfully passed quality control (*see* **Note 17** and Fig. 2).

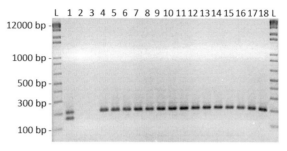

Fig 3 Duplex PCR products run on a 2 % agarose gel. *Lane 1*, input DNA; *lane 2*, PCR negative control without template; *lane 3*, MeDIP negative control without antibody; *lanes 4–18*, MeDIP samples. L, DNA ladder

3.6 Whole Genome Amplification (WGA) for Microarray Hybridization

Assuming that each MeDIP repetition will yield 150–200 ng DNA per sample, perform and pool as many repetitions as necessary to obtain enough MeDIP material for fluorochrome labeling. If this is not feasible, an aliquot of the pool can be subjected to a mild amplification using the protocol described here. For each sample, the input DNA should then be processed the same way.

1. Starting material: around 100 ng MeDIP and input DNA in 10 μL H$_2$O (*see* **Note 18**). In strip tubes, dilute 100 ng input in 10 μL H$_2$O. Add 6 μL H$_2$O to 4 μL MeDIP. Include a negative control without DNA (10 μL H$_2$O).

2. Add 2 μL 1× Library Preparation Buffer.

3. Add 1 μL Library Stabilization Solution, vortex, and centrifuge briefly.

4. Incubate at 95 °C for 2 min in a thermal cycler.

5. Immediately cool on ice and centrifuge briefly.

6. On ice, add 1 μL Library Preparation Enzyme, vortex, and centrifuge briefly.

7. Incubate in the thermal cycler as follows: 16 °C for 20 min, 24 °C for 20 min, 37 °C for 20 min, 75 °C for 5 min, 4 °C hold. Then proceed to amplification or store at –20 °C up to 3 days.

8. On ice, prepare a mix containing the following reagents per sample: 7.5 μL 10× Amplification Master Mix, 47.5 μL H$_2$O, and 5 μL WGA Polymerase.

9. Add 60 μL of the mix to the reaction, vortex, and centrifuge briefly.

10. Amplify in the thermal cycler as follows: 95 °C for 3 min, then 10 cycles of 95 °C for 15 s, 65 °C for 5 min, then 4 °C hold (*see* **Note 19**).

11. Purify using a QIAquick PCR purification kit. Elute in H$_2$O. Measure concentration and migrate 5–10 μL on a 1.5 % agarose gel (Fig. 4). Compared with unamplified samples, WGA produces a slight shift toward small molecular weights.

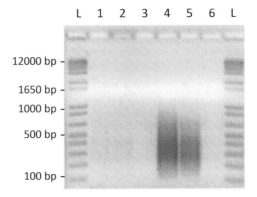

Fig 4 Migration of input (*lanes 1, 4*), MeDIP (*lanes 2, 5*) and negative control without DNA (*lanes 3, 6*) on a 1.5 % agarose gel after 5 cycles (*lanes 1–3*) and 10 cycles (*lanes 4–6*) WGA amplification. L, DNA ladder

3.7 General Considerations About Microarray Hybridization

The analysis of MeDIP samples requires tiling oligonucleotide microarrays. At present such microarrays are not commercially available for the cattle, but the two main manufacturers, Agilent Technologies and Affymetrix, propose a custom design service. DNA microarrays from Roche-NimbleGen, which have been widely used for DNA methylation analysis, are no more available.

The labeling and hybridization steps greatly depend on the microarray choice, and appropriate kits can be purchased from both manufacturers, so it would be meaningful to provide detailed protocols here. Examples can be found in [22] or [23], and comparison between the Agilent and Affymetrix technologies in [24]. Affymetrix microarrays offer the greater probe density but allow only single-channel hybridization, which implies that for the same animal the MeDIP and input samples are hybridized on two separate microarrays. Consequently, each sample must be hybridized several times to correct for the microarray effect. In contrast, Agilent microarrays offer limited density but allow two-color hybridization, so that MeDIP and input samples can be labeled with two different fluorochromes and hybridized together on the same microarray. However, dye bias is a general problem for two-color microarray experiments, so it can be beneficial to incorporate technical dye-swaps for every sample.

4 Notes

1. Given that a bovine organ can usually not be analyzed on its whole, particular attention should be paid to sampling, which should be reproducible for all animals.

2. Do not vortex, this would produce mechanic degradation of genomic DNA.

3. If the precipitate of genomic DNA is not visible, increase the amount of ethanol up to 2.5 volumes, and incubate for a few hours at −20 °C. Then pellet the genomic DNA by centrifugation.

4. Overnight incubation at 4 °C is recommended before resuspension.

5. Low molecular weight species should not be visible. Low molecular weight species can either be due to degradation of genomic DNA, or to contamination by RNA. In this case an additional RNAse treatment prior to sonication should be performed. It is important to obtain RNA-free genomic DNA, since methylation of RNAs has also been reported (reviewed in [25]).

6. We tested different DNA concentrations, and did not find that the efficiency of sonication was affected by the amount of DNA present in the solution.

7. If possible, treat all samples in parallel. We found that up to 30 samples can be manually operated without difficulties.

8. The generation of libraries for next generation sequencing of the MeDIP products requires a ligation step with adaptators on double-strand sonicated DNA. For MeDIP-seq application, ligation is directly performed on the input sample before the denaturation step. Adaptator-ligated fragments are then denatured and subjected to MeDIP prior to PCR amplification, size-selection and sequencing. For more details about MeDIP-seq, see [20].

9. Shake well the bead suspension before pipeting, in order to ensure homogenous distribution across samples.

10. The flow-through supernatant can be saved for further analysis.

11. It is important that all the samples are washed the same way.

12. Stop point: after the last washing, beads can be stored at −20 °C before to proceed to next step.

13. Be sure that the microtubes are still hermetic after several opening/closing operations. If necessary, transfer into unused microtubes before the 50 °C incubation step, to prevent leakage of the SDS-containing PK digestion buffer.

14. Caution: pellet is very loose and easy to aspirate.

15. It is not necessary to test the input DNA for each sample. One sample is enough (for instance, the sample used to generate negative controls and internal standards).

16. Migrate long enough to discriminate the two PCR products, which have very close sizes (159 and 203 bp).

17. If degradation is observed in an internal standard, do not use the samples generated during the same experiment.

18. Starting from about 100 ng DNA instead of 10 ng as suggested by the manufacturer allows to decrease the number of cycles, and therefore to limit amplification biases.

19. Starting from 100 ng, a 10 cycle-amplification will lead to approximately 3 μg DNA after purification on a QIAquick column. If less material is required, the number of cycles can be decreased. If more material is required, two 10 cycle-amplifications can be performed using the same sample and pooled.

Acknowledgements

This work has been funded by INRA (PHASE ACI 2010) and by grant ANR-09-GENM-012-01 (French National Research Agency/APIS-GENE).

The author would like to thank Sean Kennedy and the MetaQuant platform (MICALIS, INRA, Jouy-en-Josas, France) for their help with DNA sonication; Sandrine Balzergue and the transcriptomic platform (URGV, INRA, Evry, France) for their help with microarray hybridizations; Luc Jouneau (VIM/BDR, INRA, Jouy-en-Josas, France) for statistical analysis and Hélène Jammes and all team members (BDR, INRA, Jouy-en-Josas, France) for everyday support and helpful discussions.

References

1. de Montera B et al (2010) Quantification of leukocyte genomic 5-methylcytosine levels reveals epigenetic plasticity in healthy adult cloned cattle. Cell Reprogram 12(2):175–181

2. Hiendleder S et al (2004) Tissue-specific elevated genomic cytosine methylation levels are associated with an overgrowth phenotype of bovine fetuses derived by in vitro techniques. Biol Reprod 71(1):217–223

3. Couldrey C, Wells DN (2013) DNA methylation at a bovine alpha satellite I repeat CpG site during development following fertilization and somatic cell nuclear transfer. PLoS One 8(2):e55153

4. Couldrey C, Lee RS (2010) DNA methylation patterns in tissues from mid-gestation bovine foetuses produced by somatic cell nuclear transfer show subtle abnormalities in nuclear reprogramming. BMC Dev Biol 10:27

5. Cokus SJ et al (2008) Shotgun bisulphite sequencing of the Arabidopsis genome reveals DNA methylation patterning. Nature 452(7184):215–219

6. Lister R et al (2008) Highly integrated single-base resolution maps of the epigenome in Arabidopsis. Cell 133(3):523–536

7. Smith ZD, Gu H, Bock C, Gnirke A, Meissner A (2009) High-throughput bisulfite sequencing in mammalian genomes. Methods 48(3): 226–232

8. Jin SG, Kadam S, Pfeifer GP (2010) Examination of the specificity of DNA methylation profiling techniques towards 5-methylcytosine and 5-hydroxymethylcytosine. Nucleic Acids Res 38(11):e125

9. Nestor C, Ruzov A, Meehan R, Dunican D (2010) Enzymatic approaches and bisulfite sequencing cannot distinguish between 5-methylcytosine and 5-hydroxymethylcytosine in DNA. Biotechniques 48(4):317–319

10. Booth MJ et al (2012) Quantitative sequencing of 5-methylcytosine and 5-hydroxymethyl cytosine at single-base resolution. Science 336(6083):934–937

11. Yu M et al (2012) Base-resolution analysis of 5-hydroxymethylcytosine in the mammalian genome. Cell 149(6):1368–1380

12. Rauch T, Pfeifer GP (2005) Methylated-CpG island recovery assay: a new technique for the rapid detection of methylated-CpG islands in cancer. Lab Invest 85(9):1172–1180

13. Brinkman AB et al (2010) Whole-genome DNA methylation profiling using MethylCap-seq. Methods 52(3):232–236

14. Serre D, Lee BH, Ting AH (2010) MBD-isolated genome sequencing provides a high-throughput and comprehensive survey of DNA methylation in the human genome. Nucleic Acids Res 38(2):391–399

15. Mohn F, Weber M, Schubeler D, Roloff TC (2009) Methylated DNA immunoprecipitation (MeDIP). Methods Mol Biol 507:55–64

16. Weber M et al (2005) Chromosome-wide and promoter-specific analyses identify sites of differential DNA methylation in normal and transformed human cells. Nat Genet 37(8):853–862

17. Harris RA et al (2010) Comparison of sequencing-based methods to profile DNA methylation and identification of monoallelic epigenetic modifications. Nat Biotechnol 28(10):1097–1105

18. Bock C et al (2010) Quantitative comparison of genome-wide DNA methylation mapping technologies. Nat Biotechnol 28(10):1106–1114

19. Clark C et al (2012) A comparison of the whole genome approach of MeDIP-seq to the targeted approach of the Infinium HumanMethylation450 BeadChip((R)) for methylome profiling. PLoS One 7(11):e50233

20. Li N et al (2010) Whole genome DNA methylation analysis based on high throughput sequencing technology. Methods 52(3):203–212

21. Borgel J et al (2010) Targets and dynamics of promoter DNA methylation during early mouse development. Nat Genet 42(12):1093–1100

22. Zhang X et al (2006) Genome-wide high-resolution mapping and functional analysis of DNA methylation in arabidopsis. Cell 126(6):1189–1201

23. Rabinovich EI et al (2012) Global methylation patterns in idiopathic pulmonary fibrosis. PLoS One 7(4):e33770

24. Zilberman D, Henikoff S (2007) Genome-wide analysis of DNA methylation patterns. Development 134(22):3959–3965

25. Sibbritt T, Patel HR, Preiss T (2013) Mapping and significance of the mRNA methylome. Wiley Interdiscip Rev RNA 4(4):397–422

26. Gebert C et al (2009) DNA methylation in the IGF2 intragenic DMR is re-established in a sex-specific manner in bovine blastocysts after somatic cloning. Genomics 94(1):63–69

27. Gebert C et al (2006) The bovine IGF2 gene is differentially methylated in oocyte and sperm DNA. Genomics 88(2):222–229

28. Huang JM, Kim J (2009) DNA methylation analysis of the mammalian PEG3 imprinted domain. Gene 442(1–2):18–25

29. Suzuki J Jr et al (2009) In vitro culture and somatic cell nuclear transfer affect imprinting of SNRPN gene in pre- and post-implantation stages of development in cattle. BMC Dev Biol 9:9

INDEX

Nathalie Beaujean et al. (eds.), *Nuclear Reprogramming: Methods and Protocols*, Methods in Molecular Biology,
vol. 1222, DOI 10.1007/978-1-4939-1594-1, © Springer Science+Business Media New York 2015